to my friend IBrahim from Reda.
I hope that you would explore the Universe
around us using some information from
this book.
Good luck learning Arabic and other
scholastic courses.

Reda Al-Housmayne
HEATHROW Airport
06/14/92.

EXPLORING THE
Planets

Brian Jones

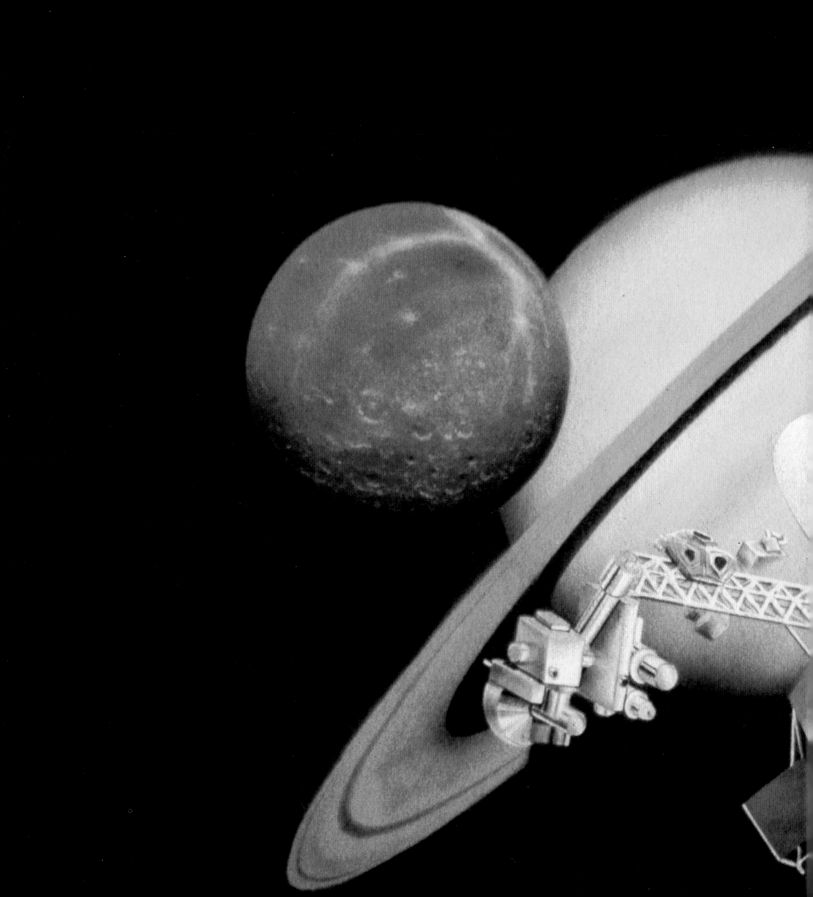

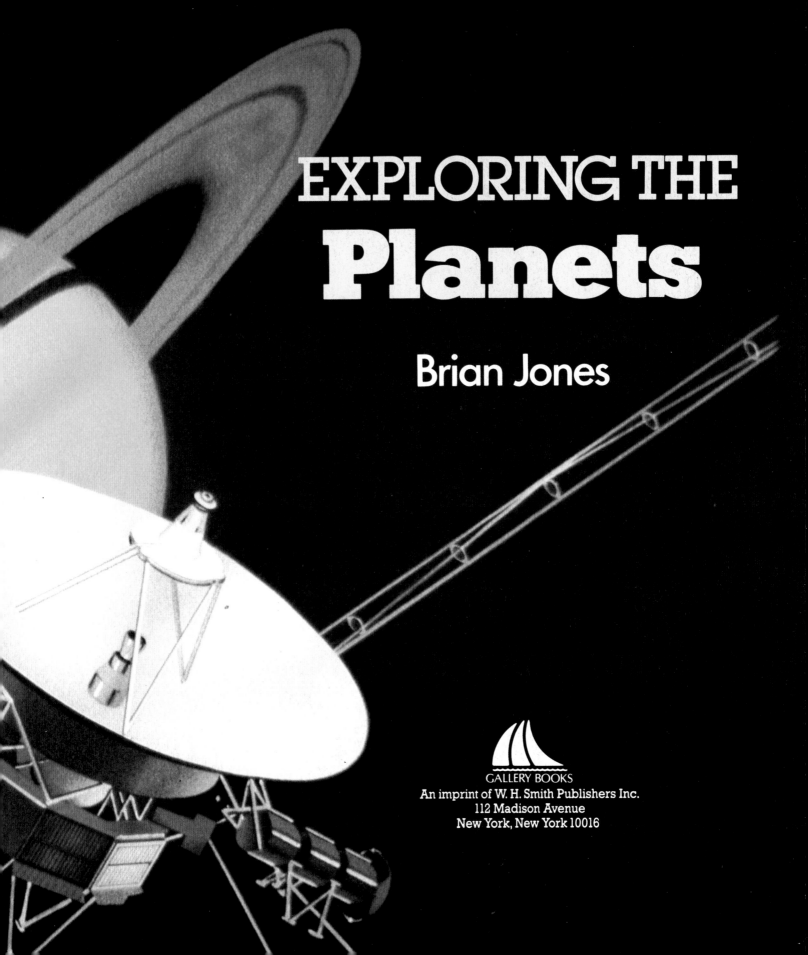

EXPLORING THE
Planets

Brian Jones

GALLERY BOOKS
An imprint of W. H. Smith Publishers Inc.
112 Madison Avenue
New York, New York 10016

First published in the
United States of America by
GALLERY BOOKS
An imprint of
W.H. Smith Publishers Inc.
112 Madison Avenue
New York, New York 10016

Copyright © 1991
Brian Trodd Publishing House
Limited

ISBN 0-8317-6975-0

Printed in Italy

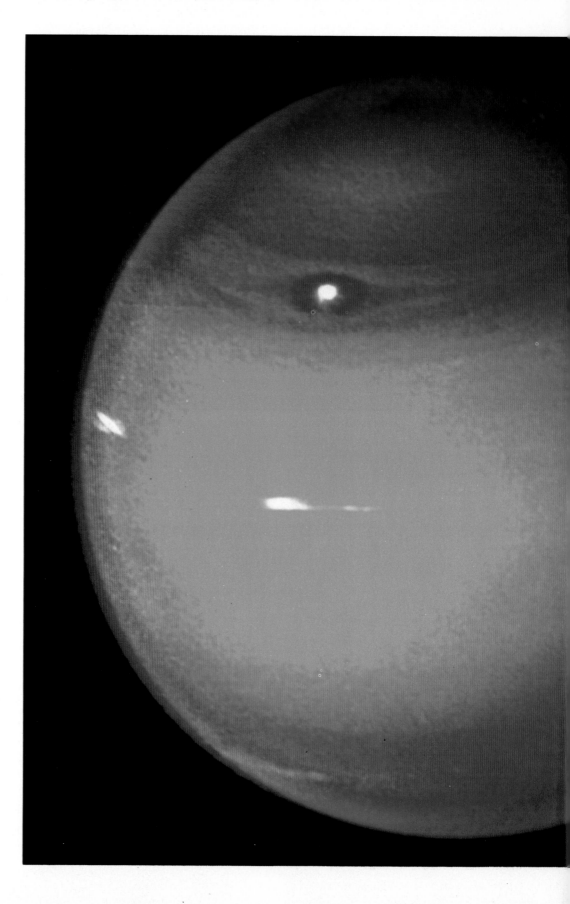

Title page: Artist Paul Doherty's
impression of the Voyager
spacecraft at Saturn.

Right: False-colour image of
Neptune taken by Voyager 2.
The red areas are
semitransparent haze covering
the planet.

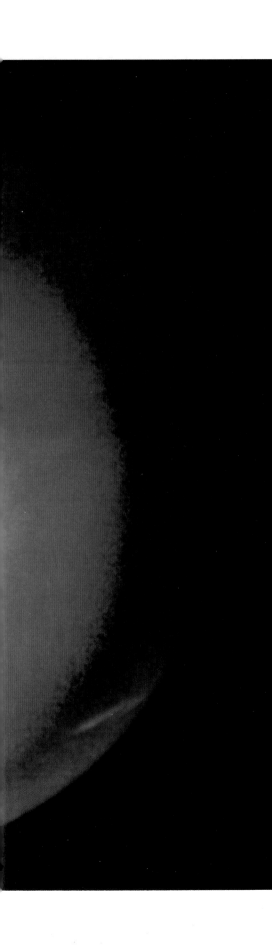

Contents

Introduction   6

Part I: The Inner Planets   11

The Formation and Evolution of the Solar System   11

The Sun   13

Mercury   17

Venus   19

Mars   22

Part II: The Asteroids   26

Part III: The Outer Planets   32

The Voyager Missions   32

Jupiter   37

Saturn   45

Uranus   50

Neptune   56

Pluto and Charon   62

Part IV: Comets and Meteors   64

Comets   64

Meteors and Meteorites   77

Part V: The Future of Planetary Exploration   80

Appendix 1   90

Appendix 2   92

Glossary   93

Index   94

Introduction

Astronomy is the oldest of the sciences, its origins reaching back almost to the dawn of civilization. The early perception of the sky was that of an inverted dome above the Earth and called the Celestial Sphere. Our visualization of the night sky today is really unchanged. The Celestial Sphere contains the stars which we see as points of light and which, quite early in history, were arranged into patterns, or groups, which we call Constellations.

Unlike the stars, which are so far away that their motions through the sky are undetectable except over prolonged periods of time, the planets and other members of the Solar System orbit the Sun, the time they take to make one orbit depending upon their distance from that star. Mercury, the innermost planet, whips around the Sun every 88 days whereas distant Pluto takes 248 years to complete one circuit.

Our perspective of the planetary motions is that the planets seem to travel through the sky, moving against the backdrop of stars. Because the planets all orbit the Sun in more or less the same plane, they all appear to keep within a particular band of sky that we call the Zodiac. The Zodiac is divided into twelve different constellations and to many ancient peoples the positions of the planets within the Zodiac, and the position of the Sun (which also keeps to this band of sky) at the time of a person's birth, were held to have effects on that person's destiny. Indeed, many of the

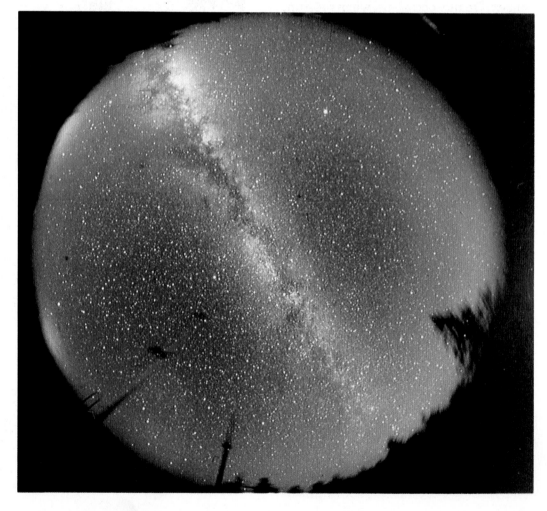

The Celestial Sphere circa 1990. A fish-eye lens view of the Milky Way and neighbouring sky.

astronomers of ancient times were actually observing the sky for astrological purposes, astrology being the art of forecasting earthly events from the study of the motions of the Sun and planets through the sky.

The stars on the Celestial Sphere all appear to lie at the same distance from us, although this is merely a line of sight effect. In reality, some stars are at vastly greater distances than others, although their apparent suspension on the celestial sphere gives them the appearance of being at equal distances.

Our position in space, on the third planet out from the Sun, dictates how we see the other planets move through the sky. The inner planets (Mercury and Venus) are always seen close to the Sun in the sky. They undergo a series of changing positions relative to the Sun as we see them, sometimes being visible in the east before sunrise or in the west after sunset. On rare occasions, when the Sun, the Earth and one of the inner planets are exactly aligned, we see the planet cross the solar disc. Such an event is known as a 'transit', the planet being seen as a tiny black speck against the brilliant solar background.

The outer planets appear to move in a different way. Mars, the next planet out from the Sun after Earth, moves fairly quickly, although the outer planets Uranus and Neptune, with their slower orbital speeds, seem to take much longer. Pluto, the outermost planet, has a motion which is difficult to detect at all except over a period of several days. The efforts of Clyde Tombaugh (see 'Pluto and Charon') testify to this.

Early Western and Arab civilizations believed that the Earth was at the centre of a spherically-shaped universe with the Sun, Moon and the then known planets orbiting around it. This notion had its difficulties, a typical example being retrograde, or backwards, motion often displayed by the planets Mars, Jupiter and Saturn (Uranus, Neptune and Pluto weren't known at that time). This motion was simply an effect of perspective which appeared as the Earth, because of its quicker motion, 'overtook' the planet in question. In order to explain the *observed* motions of the planets, an elaborate system of orbits within orbits was postu-

lated by Ptolemy, a Greek-Egyptian scholar who lived *c.* A.D. 150. Ptolemy propounded his ideas in a book called the *Almagest*, a tour de force of Greek astronomy. The Ptolemaic system was accepted for over 1300 years.

It was not until the 16th century that the first authoritative challenge to the Ptolemaic Theory was put forward. The Polish astronomer Copernicus suggested that it was the Sun, and not the Earth, that occupied the central position. The planets, he said, all travelled around the Sun in circular orbits. Needless to say, his idea, which became known as the Copernican Theory, received little initial support.

A 16th century engraving of Claudius Ptolemy (A.D. c.100–170) using a quadrant. The Ptolemaic system held that the Earth was at the centre of the universe.

Johann Kepler (1571–1630). Kepler formulated the three fundamental laws of planetary motion based on Tycho Brahe's observations.

The Danish astronomer Tycho Brahe was probably the greatest naked eye observer of all time. During the latter part of the 16th century he made a long series of observations of planetary motions, while still firmly believing in the old Ptolemaic Theory. However, it was Brahe's observations that were to provide the turning point in our understanding of planetary motion.

Brahe's assistant, a brilliant German astronomer and mathematician called Johannes Kepler, believed in the Copernican Theory, although he noticed that the *observed* motions of the planets did not fit in with the idea of circular orbits. After a few years work, following the death of Tycho Brahe and using Brahe's observations, he finally realized that the planets did indeed orbit the Sun, but in paths that were *elliptical* and not circular. At last, the true layout of the Solar System had been determined and the motions of the planets properly defined.

With the invention of the telescope in 1608 and the subsequent work of Galileo Galilei (1564–1642), the Greek notion that the other celestial bodies were featureless globes was disproved and it was realised that the other planets were true worlds in their own right.

Even with bigger and better telescopes our knowledge of the other planets remained largely incomplete and it was left to science fiction writers to paint pictures of these worlds.

The Space Age has revealed that the Earth's planetary neighbours are fascinating worlds. Ranging from the barren and crater-strewn landscapes of Mercury out to the exciting surface of Neptune's satellite Triton, astronomers and scientists have now much to ponder upon as they try and piece together the history of the Solar System. The images obtained courtesy of the Mariner, Venera, and Voyager probes, and others, have shown everything from the towering cliffs and ravines of Miranda to the deep canyons of Mars and Venus, from

Representation of the Ptolemaic system. It was assumed that because the circle is the 'perfect' form and only perfection is allowed in the heavens, then all celestial orbits must be perfectly circular.

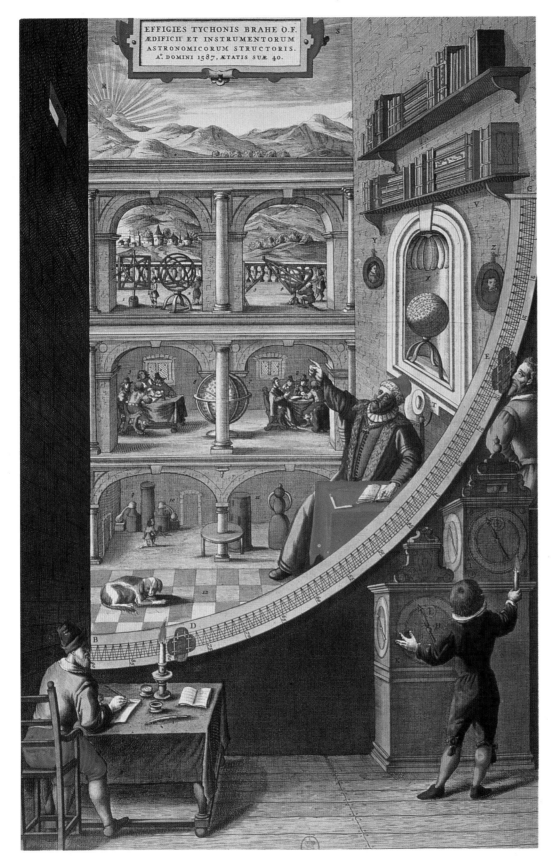

Engraving depicting the Danish astronomer Tycho Brahe (1546–1601) in his observatory on the island of Ven, Sweden. He was the last and greatest of the naked-eye astronomers.

Painting showing NASA's Voyager 2 as it might have appeared when it encountered Uranus on 24 January 1986. The missions of Voyagers 1 and 2 represented the greatest triumph of astronomy since the invention of the telescope.

active volcanoes scattered across lively Io to the dormant giant Olympus Mons whose fiery output helped to mould the Martian surface we see today. Even comets have now been explored by space probe, notably Halley's Comet during its return in 1985–86.

Following the encounter with Neptune by Voyager 2 in 1989, now only dim and distant Pluto awaits our close attentions. We stand on the threshold of the next phase of planetary exploration. Many exciting new missions are cur-

rently under way and many more are being planned. The Saturnian system, in particular its fascinating satellite Titan, will fall under the watchful of the Cassini mission, while that of Jupiter and its main satellites will by then have been closely scrutinized by the Galileo probe, already on its way to a rendezvous with the giant planet. One thing we can be sure of – every exciting discovery simply opens the door to yet more mysteries and discoveries.

Part I
The Inner Planets

The Formation and Evolution of the Solar System

Our Solar System is at least 4.7 thousand million years old, this figure being the best available estimate we have for the age of the Sun, whose birth preceded the formation of the planets. The processes by which stars such as our Sun are born are not well understood as yet, but there seems little doubt that the material from which it was formed was one of the clouds of interstellar gas and dust found throughout the Universe. In order for a star to form from such a cloud, the dust and gas must become denser by a process of accretion or by collapse inwards. Quite what caused the cloud in our region of space to increase in density is not known. It is possible that the cloud accreted matter over a long period of time until it was sufficiently large that it began to collapse in on itself due to self-generated gravitational forces. Alternatively, thermal pressures from nearby young, hot stars or from a supernova may have initiated a process whereby the interstellar dust began to clump together, forming regions of relatively high density; gravitational collapse then completed the process.

Whatever the cause, as the cloud collapsed it would have begun to heat up and a *protostar* begun to form. Such stars can be observed now in the regions of our Galaxy where there are high concentrations of dust and gas.

The evolution of a protostar into a fully-fledged star, or what are called main sequence stars, can take hundreds of thousands or even millions of years. A protostar continues to collapse under gravity until its centre reaches about 10 million °C, at which temperature nuclear fusion can begin. In such a process, identical to that which takes place in a hydrogen bomb, nuclei of hydrogen fuse

The comparative sizes of the planets. From the Sun outwards – Mercury, Venus, Earth, Mars, Jupiter, Saturn, Uranus, Neptune, Pluto.

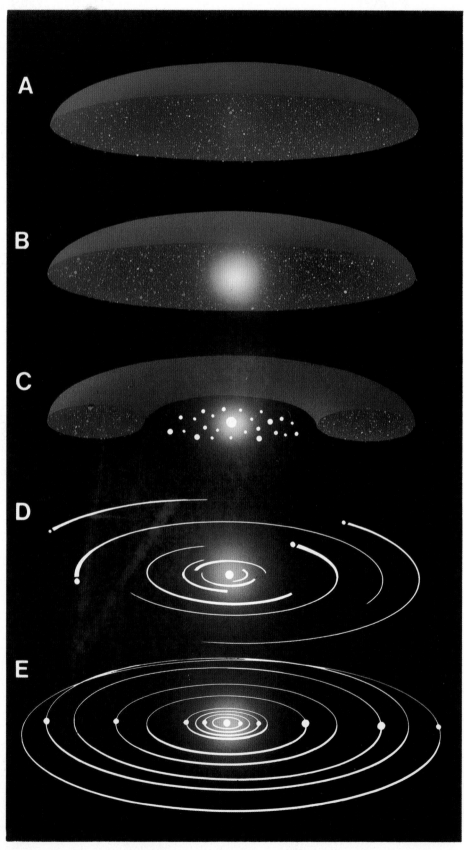

together to form helium and energy is released. The process becomes continuous and the star attains a stable structure with inward-directed gravitational forces being balanced by outward-pushing thermal pressures. The star is now a main sequence star, meaning that it has entered the longest phase of its life. Our own Sun has such a life of some ten thousand million years.

Planetary Formation

In order to trace the evolution of the planets we must return to the protostar stage. Such a star rotates, as a result of which its outer regions settle into a spinning, flattened disc. Such a disc contains particles ranging in size from microscopic up to about a kilometre. From these fragments the planets formed.

As the newly-formed star began to radiate energy, less heavy molecules near the centre of the disc would have been driven towards the farther reaches of the nebula. As this happened the refractory molecules, such as iron and the silicates condensed out nearer the star. In this way, the basic compositional differences between the inner and the outer planets of the Solar System can be readily understood. There is good reason to believe that planetary systems when they form elsewhere will not be dissimilar to our Solar System.

It is thought that the planets were gradually built by a process of collision and coalescence of the condensed particles in the planetary nebula. The early results of this process are called planetisimals, which in turn by further collision and coalescence became planets. The force of gravity aided this process as time went by, as well as causing the new planets to adopt orbits which were regular and well-spaced.

At this early stage, the inner, terrestrial planets were molten, primarily as a result of the heat generated by radioactive decay. This allowed the heavier materials within them, such as iron, to sink towards the central regions of the new planet, leaving the less dense materials nearer the surface.

The outer, gas giants of the Solar System formed in a similar way. Once they had attained a mass similar to that of the Earth they would have had sufficient gravity to collect in the hydrogen and helium that had been driven to the outer regions of the nebula by the newly-formed star at its centre, as well as any other gaseous components of the cloud.

The Sun, meanwhile, was gradually becoming hotter until its core reached a temperature perhaps as high as 15 million °C. The evolution of the Solar System was well under way.

The Sun

Despite its appearance and importance to us here on Earth, the Sun is in fact a fairly ordinary star by Galactic standards. It is classed as a yellow dwarf, the colour referring to its luminosity and temperature on a scale which runs from blue (very hot) to dim red. It has a diameter of 1,392,530km (865,300 miles) and contains 98 per cent of the entire mass of the Solar System. The activity of the Sun is constantly monitored as astronomers seek to gain a better understanding of stellar behaviour.

Sunspots

The most prominent type of feature visible on the outer layer of the Sun, or photosphere, are sunspots, dark patches which are areas some 1,500 °C cooler than the surrounding region. The size of sunspots varies considerably from some 1,000km (600 miles) to over 40,000km (25,000 miles). Their lifetime seems to depend on their size, with small spots lasting only several hours, while the largest spots may persist for weeks or even months.

Sunspots are characterized by a dark central region known as the umbra, which in turn is surrounded by a greyer region, the penumbra.

Sunspots are caused by strong magnetic fields produced in the Sun's interior. When the magnetic field lines break through the surface they cause localized cooling by suppressing the normal convection currents rising from the lower regions of the Sun. The strength of the magnetic fields around sunspots is now known to be thousands of times stronger than the Earth's magnetic field.

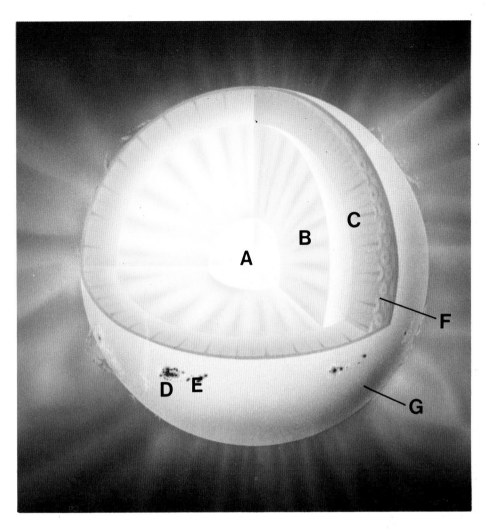

While sunspots can appear at any time, there is a definite cycle of sunspot activity. This cycle was first noted by Heinrich S. Schwabe, a German apothecary and amateur solar observer, who made his discovery public in 1843. Schwabe announced that the number of dark spots visible on the solar surface seemed to vary in a regular, roughly 10-year cycle. It was the Swiss astronomer J. Rudolf Wolf who later (1855) discovered the true figure to be closer to 11 years. Writing in 1990, the Sun is now near to one of its peaks of sunspot activity.

This variation of sunspot number is now known to be the most visible aspect of a profound oscillation of the Sun's magnetic field that affects other aspects of both the surface and the interior.

Observations of the sunspot cycle show that during the early part of the cycle the spots appear some 30 degrees

Structure of the Sun, A – core; B – radiative zone; C – convection zone; D – sunspot/active region; E – surface rotation; F – giant cell; G – supergranulation.

Left: Theory of the Formation of the Solar System. The Solar System began as a spinning cloud of gas and solid grains of ice and rock (A). As the cloud began to collapse under its own gravity, a region of greater density at the centre became a proto-Sun, which, when it heated up, began to evaporate the ice in the regions nearest to it (B). Nearer the Sun the dust particles condensed to form rocky planetesimals, while further out the planetesimals also incorporated ice – hence the reason for the division in composition between inner and outer planets (C). Later, the rocky planetesimals collided and coalesced to form planets, while the larger, outer planetesimals attracted more gas, becoming gas giants (D, E).

north or south of the equator. As the cycle progresses sunspots develop at lower and lower latitudes until at the end of the cycle they are seen very close to the equator.

Observations of sunspots have helped astronomers to determine the axial rotation of the Sun, which has been found to vary depending on the latitude. Work done by the British astronomer Richard Carrington in 1859 showed that the Sun does not rotate as a solid body, a phenomenon which is given the name differential rotation. A sunspot at the equator is carried around the Sun in about 25 days; those at greater latitudes take longer. The rotation period at the solar poles is thought to be as long as 35 days. This differential rotation is possibly the key to the production and maintenance of the Sun's enormous magnetic field, by producing an effect similar to that of a dynamo.

Another visible phenomenon of the photosphere also associated with magnetic disturbances are *faculae*, irregular patches or streaks brighter than surrounding areas and which typically occur in the neighbourhood of sunspots. They are in fact clouds of incandescent gas in the upper regions of the photosphere. Such clouds often precede the appearance of sunspots.

Sunspots appear to be regions of the solar surface that have been pierced by magnetic loops from the interior. The uniform magnetic field below the solar surface gets tangled *(right, top)* by the turbulent gas in the convection zone. The field bursts through the photosphere *(right, below)* and forms a sunspot pair.

Energy generated in the core of the Sun is carried upwards by two means, radiation and convection. Near to the core, radiation disperses energy, while from about 0.85 solar radii up to the photosphere convection currents are responsible for energy transport. From the surface outwards, radiation again takes over.

Observation of the Sun's surface shows that the photosphere is not uniformly bright but has a mottled texture called *granulation*. The smallest granules consist of bright patches of light about 1,000km (620 miles) across with a dark border. The pattern of granules changes continuously over a period of minutes. Measurements show that the bright centre of a granule is moving upward, whereas the boundary is a cooler, descending gas. Almost certainly granules are the highest tier in what are thought to be three convection zones reaching down towards the core, each containing 'cells' of convection. The deepest layer contains the giant cells, each possibly encompassing 200,000km (124,300 miles) while above this are the super-granular cells, perhaps 30,000km (18,600 miles) in diameter. Above these, roughly 2,000km (1,200 miles) deep and reaching to the surface is a layer of smaller convection current cells. The tops of this layer make up the photosphere and are responsible for the granulated appearance.

The Chromosphere and Corona

Above the photosphere lies the chromosphere, the Sun's atmosphere. It consists largely of hydrogen, helium and calcium and is usually only visible at the time of a solar eclipse. Studies have shown that it extends some 10,000km (6,000 miles) above the photosphere and merges into the corona which consists of ionized gas, or plasma. The temperature of the chromosphere increases with height, being some 4,500°C at the bottom and rising to 100,000°C at the top. The temperature of the corona exceeds 1,000,000°C.

Two phenomena of the Sun's atmosphere are flares and prominences, both associated with magnetic disturbances. Flares are sudden, violent releases of energy associated with sunspot groups. They are only occasionally seen in visible light but more often they emit radiation in the form of X-rays and the extreme ultraviolet. Flares usually last from a few minutes to a few hours; large masses of plasma may be ejected into space. They are sometimes so violent that they cause additional ionization in the Earth's ionosphere and may disrupt radio transmissions.

Prominences are luminous clouds of gas that appear in the corona and can be quite spectacular, particularly when viewed during a solar eclipse. It was thought by early solar observers that prominences were the result of matter being thrown out of the photosphere, but it is now realized that they usually result from matter which has condensed in the corona, descending. There are two types of prominence, quiescent and active or eruptive. The former last for several solar rotations and dissolve slowly. The latter rise rapidly and violently, have complex forms and last for less than an hour or so. They can reach heights of 700,000km (435,000 miles). Some prominences are known as loop prominences in which condensing matter returns to the photosphere along a loop of magnetic flux.

The last sphere of solar activity is the heliosphere, the region of space in which the solar wind is dominant. The helio-sphere is thought to extend out to a distance of almost 15 thousand million kilometres, at which point the Sun's influence gives way to interstellar space. The boundary of the heliosphere marks the true outer limit of the Solar System.

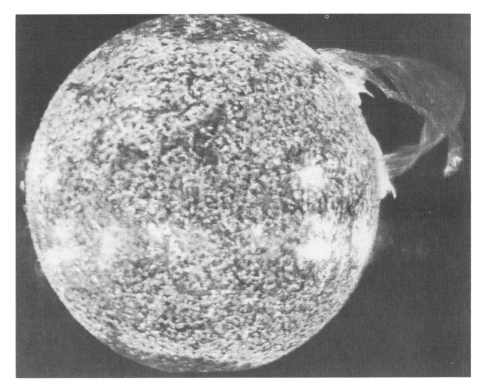

One of the most spectacular solar flares ever recorded and spanning more than 588,000km (367,000 miles) across the solar surface. Taken by Skylab, 19 December 1973.

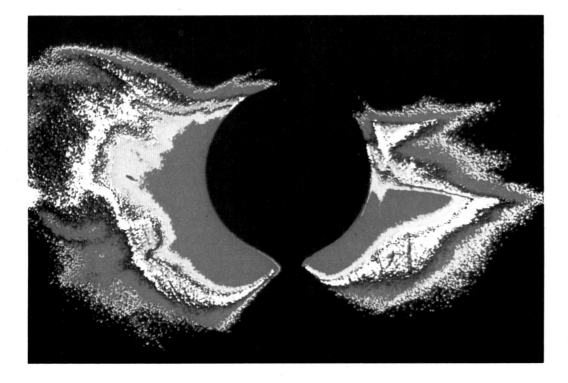

A colour-enhanced image of the solar corona, taken by Skylab.

The Promise of Ulysses

Until now, all our observations of the Sun have been carried out either from the Earth's surface or from spacecraft operating within or close to the ecliptic, the plane of the Earth's orbit around the Sun. In other words, we have been unable to observe the Sun from above its polar regions.

The joint NASA/ESA Ulysses mission promises to change this. Originally known as the 'Out of Ecliptic' mission, and then as the International Solar Polar Mission (ISPM), the project has been named Ulysses after the mythological Greek hero of the Trojan War and King of Ithaca described in Dante's Inferno. We are told that Ulysses decided to '. . . venture the uncharted distances . . . of the uninhabited world behind the Sun . . . to follow after knowledge and excellence'.

The Ulysses mission, due for launch from the Space Shuttle towards the end of 1990, will allow us to greatly expand our knowledge of the Sun. The prime objective of the Ulysses mission is to explore the inner heliosphere over the full range of solar latitudes, thereby allowing scientists to make substantial progress in our understanding of the physics of the heliosphere. Following launch, the probe will make its way out to Jupiter, which it will encounter 16 months later. During its flyby of Jupiter, the probe will use the planet's immense gravitational pull as a slingshot to propel the craft out of the plane of the ecliptic and arcing back towards the Sun. It will pass over the south solar pole in 1994 and then move north across the ecliptic for a pass over the north solar pole a year later. After many delays, primarily caused by the suspension of Shuttle flights following the Challenger accident in January 1986, the Ulysses mission promises to provide astronomers with a wealth of new information about our parent star and its influence on our neighbourhood.

An artist's impression of the route the spacecraft Ulysses will take on its way to the south and north pole of the Sun. The fly-by of Jupiter occurs 15 months after launch.

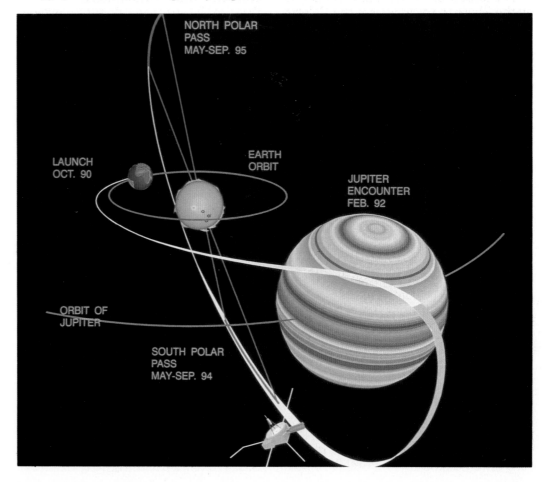

Mercury

Mercury is the innermost planetary member of the solar system, orbiting the Sun once every 87.97 days at a mean distance of 58 million kilometres (36 million miles). It is similar in size to the Moon, with a diameter of only 4,880km (3,032 miles). Its similarity to the Moon does not end with its diameter, the cameras on board the Mariner 10 spacecraft, launched in November 1973, revealing a surface covered in craters, mountains, ridges and valleys. Mariner 10 made the first ever flyby of Mercury by a space probe in March 1974, this being followed by further passes in September 1974 and March 1975. Although craters were in abundance, there were fewer of the dark plains that dominate the lunar surface. The success of Mariner 10 has resulted in much of Mercury's surface having now been mapped, providing us with valuable information.

The largest of the dark basins imaged by the Mariner 10 cameras is the Caloris Basin. This is a huge impact formation, some 1,300km (807 miles) in diameter and which is surrounded by mountain blocks reaching up to 2km (1.2 miles) above the surrounding plains. Only half the Caloris Basin was photographed because the same areas of the Mercurian surface were presented to Mariner 10 during each pass.

The destructive effects of the impact event which produced the Caloris Basin were not confined to the impact site. At the antipodes of the Caloris Basin (the point on the other side of the planet directly opposite the impact site) is the 170 kilometre diameter crater Petrarch. Around this crater numerous fractures formed as a result of the seismic shocks set up by the impact. These shock waves were focused through the planet towards the antipodes. One of these fractures is the Vallis Arecibo, a 7km (4.3 miles) wide valley which stretches between Petrarch and the 80km (49.7 miles) diameter crater Ibsen, which lies around 100km (62 miles) to the north-west.

Despite Mercury being only one-third the diameter of the Earth, the density of the two planets is similar. From this it is inferred that the planet's core is largely composed of iron.

This photograph of Mercury was taken from 200,000km (125,000 miles) by Mariner 10 in 1974.

Rotation and Atmosphere

Prior to the space age all our knowledge of Mercury came from Earth-based observations, the first serious attempts to map the planet being made by Giovanni Virginio Schiaparelli during the 1880s. His rough chart contained nothing more than light and dark areas and told us little about the planet. A more detailed chart was produced by Eugenios Antoniadi between 1924 and 1933, although his observations have since been found to be inaccurate.

Both Schiaparelli and Antoniadi believed that Mercury had a captured, or synchronous, rotation period. This would mean that Mercury's orbital period and axial rotation would be 88 Earth-days, with one hemisphere permanently facing the Sun and being subjected to everlasting day while the other would be in permanent darkness. The eccentricity of Mercury's orbit would result in a 'twilight zone' between the two extremes of day and night, where the Sun would rise and set, always staying close to the horizon. The effects would be similar to lunar libration. However, radar measurements carried out in the early 1960s showed that the true axial rotation period was 58.65 days, thereby ensuring that all parts of Mercury's surface receive sunlight at some time or another.

Mercury's Caloris Basin can be clearly seen in this enhanced Mariner 10 photo. A 1,300km (800 mile) diameter ring of mountains up to 2km (6,500ft) high rims the outer edge of the basin.

One curious aspect of the axial rotation period is that the planet spins exactly three times during two orbits of the Sun. The result of this is that the same hemisphere is presented to Earth every time the planet is best placed for observation. It is this fact that led early observers to believe that Mercury had a captured rotation. This so-called spin-orbit coupling also means that the Mercurian solar day (sunrise to sunrise) is 176 Earth-days long, or two Mercurian years.

Antoniadi also mistakenly believed that Mercury had an atmosphere, due to his supposed observations of clouds above the surface. We now know that the Mercurian atmosphere is far too tenuous to support clouds. A major contributor to the lack of an atmosphere around the planet is the fact that Mercury's escape velocity is only 4.3 km/sec (2.7 miles/sec). Mariner 10 instruments detected traces of hydrogen and helium near the surface, probably originating from the Sun. Spectroscopic observations made from Earth in 1985 also led to the discovery of sodium in Mercury's atmosphere. The element appears to be the most abundant component of an atmosphere so rarified that it betters the best vacuum capable of being produced in a laboratory on Earth. The virtual lack of an insulating atmosphere largely accounts for the wide range in Mercurian surface temperatures, which can be as high as 425°C on the equator at noon, plummeting to −180°C just before sunrise. Such a hostile environment may preclude manned landings.

Venus

Venus orbits the Sun once every 224.7 days at a mean distance of just over 108 million km (67.1 million miles) It is the closest planet to Earth and the brightest object in the sky (apart from the Sun and Moon) and has even been known to cast shadows. This brilliance is due to its covering of dense clouds which reflect over three-quarters of the sunlight received by the planet. Although giving Venus a visually stunning appearance, these clouds completely hide the surface of the planet from view and Earth-based observations of Venus are limited to the cloud tops.

Rotation and Atmosphere

Although there are atmospheric markings (first detected by the German astronomer Johann Hieronymus Schröter in 1788) the visible disc contains no features permanent enough to allow accurate measurements of the rotation period to be made, and prior to the space age and the advent of powerful radar, this remained unknown. The first radar observations of the planet were made in 1961. Radar signals were bounced from the surface of the planet, those received back from the approaching side being of a higher frequency than those from the receding side due to the Doppler Effect. Together with subsequent observation, astronomers have now deduced that the axial rotation is retrograde (in a direction that is opposite to its orbital motion) with a period of 243.01 days. Like Mercury, Venus has a 'day' that is longer than its 'year'.

While appearing beautiful, the clouds conceal a hostile atmosphere, the main constituent of which is carbon dioxide. However, traces of many other substances have been detected including hydrogen sulphide, carbon monoxide, sulphur dioxide and hydrochloric acid. The atmosphere reaches to a height of around 250km (155 miles) with over 90 per cent of it being concentrated within 30km (18.6 miles) of the surface. This has produced a surface pressure equivalent to around 90 times the atmospheric pressure at the Earth's surface.

Not only is the atmospheric pressure extremely high. The high carbon dioxide

Venus taken with an ultraviolet filter from Mariner 10. The blue coloration was added to show the circulation pattern of the clouds.

Four views of Venus taken by Pioneer Venus Orbiter several days apart. In picture No. 2 a dark, Y-shaped feature, often seen in Venus' clouds is visible. It is thought it may be sulphur clouds lying well below the highest cloud layer.

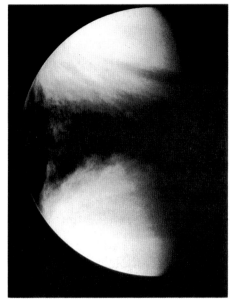

No. 1—Dec. 30, 1978

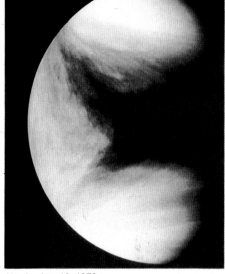

No. 2—Jan. 10, 1979

No. 3—Jan. 14, 1979

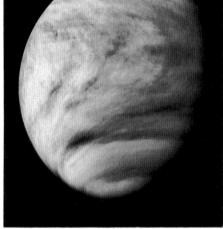

No. 4—Jan. 18, 1979

Venera 4 successfully descended onto the surface of Venus – a considerable feat in view of the hellish conditions there. This illustration shows how the landing was achieved.

content of the atmosphere has produced a runaway greenhouse effect, with the heat from the Sun that filters down to the surface becoming trapped and unable to escape back into space. The temperature at the Venusian surface is round 450°C.

Soviet and American space probes have told us all we know about the Venusian surface. Prior to exploration by space probe, there were several theories as to what the surface of the planet might be like under the dense cloud cover, including the idea that the surface of the planet was entirely covered by water.

But what does the surface look like? The Soviet Venera 9 craft, which landed on a slope near a hill in the vicinity of Beta Regio (see below), on Venus in October 1975, sent back a single, wide-field black and white picture which revealed a rocky terrain with stones several tens of centimetres across and soil scattered between them. The temperature at the landing site was measured at 460°C and the windspeeds no higher than 2.5km (1.5 mph) per hour.

The Venusian Topography

While Soviet craft provided us with pictures of the Venusian surface, further probes carried out mapping of the planet. Unlike other planets that can be mapped with straightforward imagery, such as Mercury or Mars, the dense cloud covering of Venus poses problems. A special technique known as radar mapping has been used to provide information on the Venusian topography.

Radar mapping has been carried out both by Soviet and American spacecraft and involves an orbiting vehicle which transmits radar pulses to the surface. The time elapsed between the pulse being transmitted and being reflected back to the waiting spacecraft is carefully measured. The longer a radar pulse takes to return, the further it has had to travel, those hitting valleys and other low-lying areas taking longer to return than those being reflected from mountains and other highland regions.

Radar mapping has revealed that almost three-quarters of the Venusian surface is comprised of flat rolling plains, although there are several highland regions including two named Aphrodite Terra and Ishtar Terra. Aphrodite Terra straddles the Venusian equator and is the larger of the two, being comparable to Africa in size. Ishtar Terra is found further to the north and is roughly the size of Australia.

Ishtar Terra has a mean height of around 3km (1.86 miles) and contains several mountains including Maxwell

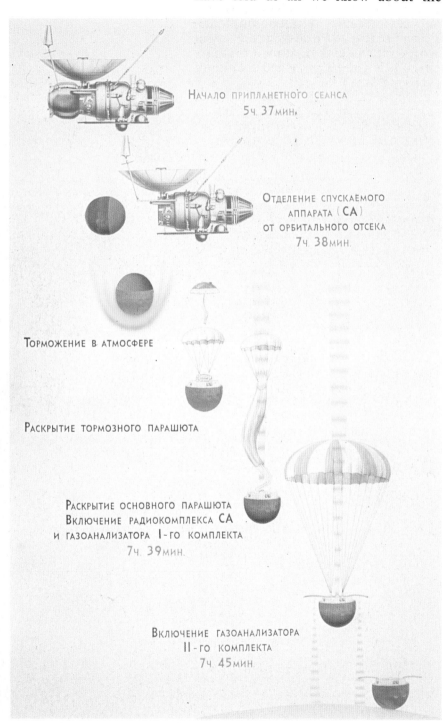

НАЧАЛО ПРИПЛАНЕТНОГО СЕАНСА
5ч. 37мин.

ОТДЕЛЕНИЕ СПУСКАЕМОГО АППАРАТА (СА)
ОТ ОРБИТАЛЬНОГО ОТСЕКА
7ч. 38мин.

ТОРМОЖЕНИЕ В АТМОСФЕРЕ

РАСКРЫТИЕ ТОРМОЗНОГО ПАРАШЮТА

РАСКРЫТИЕ ОСНОВНОГО ПАРАШЮТА
ВКЛЮЧЕНИЕ РАДИОКОМПЛЕКСА СА
И ГАЗОАНАЛИЗАТОРА I-го КОМПЛЕКТА
7ч. 39мин.

ВКЛЮЧЕНИЕ ГАЗОАНАЛИЗАТОРА
II-го КОМПЛЕКТА
7ч. 45мин.

Montes, the highest mountains on Venus, situated near the centre of Ishtar and rising to around 11km (6.8 miles) above the mean surface level. Near Maxwell Montes is Cleopatra Patera, a 1.5km (0.9 miles) deep, 100km (62 miles) diameter impact crater with a small crater 1km (0.6 miles) deep and 55km (34 miles) across at its centre.

Another highland region is Beta Regio which contains Rhea Mons and Theia Mons, two large and possibly-active shield volcanoes which tower 4km (2.5 miles) high. Another feature of note is Diana Chasma, a huge valley comparable in size to Vallis Marineris on Mars. Diana Chasma is the deepest fracture on Venus with a depth of around 2km (1.2 miles) below the surface and a width of nearly 300km (186 miles).

Despite the fact that Venus possesses a large nickel-iron core, it does not have a magnetic field, the slow rotation of the planet making it unable to generate a field by the usual dynamo action.

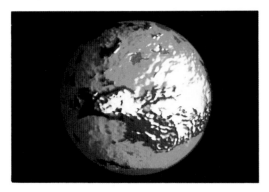

Left: Sophisticated computer graphics transformed radar data from Pioneer Venus Orbiter to this image of Venus showing the surface features hidden beneath the clouds. The prominent feature near the centre of the globe is the highland region of Aphrodite Terra.

The highest and most dramatic continent-sized highland region of Venus is Ishtar Terra, shown here in an artist's conception based on topography measurements by Pioneer Venus Orbiter. An outline of the USA is included for comparison.

Mars

Mars is the outermost of the terrestrial planets, travelling around the Sun every 687 days in a somewhat eccentric orbit that takes it between 206.7 million and 249.1 million kilometres (128.4–154.7 million miles) from our star. Often called the Red Planet, its conspicuous tint arises from large amounts of reddish dust scattered across the Martian terrain and which led to the planet being named after the legendary God of War. Winds that sweep across the planet can raise huge volumes of this dust high into the thin Martian atmosphere, temporarily hiding large areas of the Martian surface from view.

The attentions of many astronomers have been drawn to the Red Planet, one of the earliest being the Italian Giovanni Domenico Cassini who, in 1666, made the first reasonably accurate measurements of Mars' axial rotation period, which he found to be $37\frac{1}{2}$ minutes longer than that of the Earth (the actual difference is just over 41 minutes). Cassini was also the first to observe the Martian polar ice caps, although it was to be more than a century before William Herschel first suggested that these may contain water ice.

The first observations of surface mark-

The approach to Mars by Viking 1 in June 1976. After arriving in Mars orbit, the Viking lander was despatched to the surface.

ings were made in 1659 by Christiaan Huygens who drew the dark triangular feature we now know as the large plateau Syrtis Major. The first comprehensive charts of Mars were compiled by Wilhelm Beer and Johann von Madler during the 1830s.

One of the most famous observers of Mars was the American Percival Lowell. In 1877 the Italian astronomer Giovanni Schiaparelli reported the observation of a number of linear features crossing the Martian surface. He referred to them as 'canali', an Italian expression meaning natural channels. However, the word was erroneously translated as 'canals', which are anything but natural.

Soon the idea of an intelligent Martian civilization took hold and astronomers, notably Lowell, began to report the existence of more Martian canals, Lowell himself listing over 150. He believed that Mars was a desert world, the inhabitants of which were irrigating vegetation near the equator by using canals to convey water from the ice caps. Exploration of Mars by space probe has finally disproved their existence.

The first successful probe to Mars was the American Mariner 4, which flew past Mars in July 1965 sending back the first close-up pictures of the Martian surface. Lowell's canals were conspicuous by their absence. Numerous other features were seen, however, including large numbers of craters. During the 1890s the American astronomer Edward Emerson Barnard had reported seeing craters on Mars, although his results were never published, presumably for fear of ridicule.

Mariners 6 and 7 (1969) and Mariner 9 (1971–72) carried out successful follow-up missions, although it was the American Viking 1 and Viking 2 missions (1976) which gave us much of our present day knowledge of the planet. Each of the Viking craft comprised an orbiter and a lander, the landers sending back a great deal of information including details of surface changes, meteorology and the Martian atmosphere. The pictures received from the Viking landers showed that the sky had a reddish tint, due to fine dust being suspended in the atmosphere. The atmospheric pressure was found to be less than 1 per cent of that at the Earth's surface, the atmosphere itself consisting mainly of carbon dioxide with smaller amounts of nitrogen, argon, oxygen, carbon monoxide and water vapour.

The Martian Terrain

Although Mars has long been a target for Earth-based astronomers, relatively little was known about the Martian surface prior to the Space Age. The probes that have explored the planet have revealed a landscape containing many prominent features. Notable examples are the impressive group of shield volcanoes on the huge bulge of the Tharsis region, the largest of which is Olympus Mons. This is by far the most visually stunning volcano in the Solar System, towering around 25km (15.5 miles) above the surrounding plains. Its base diameter is over 600km (372.8 miles) while the caldera is 80km (49.7 miles) across. Its volume exceeds even the largest terres-

Bottom: A frosty scene near Mars' north pole showing the region in mid-summer when the seasonal carbon dioxide polar cap clears to reveal water ice and layered terrain beneath.

Below: A high resolution, colour-enhanced image of the Candor Chasm in the Vallis Marineris on Mars, taken by Viking Orbiter 1, 1976.

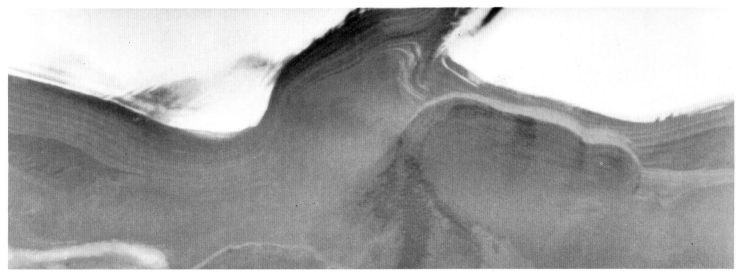

An artist's impression of a future base on Deimos.

In 1738, the French writer Voltaire wrote a short science fiction tale (eventually published in 1752), originally titled *The Adventures of Baron Gangan* although later retitled *Micromegas*. In this story he described the possible existence of '. . . two moons subservient to that orb (Mars) which have escaped the attentions of . . . astronomers'. He stated that '. . . Mars, which is at such a distance from the Sun, must be in a very uncomfortable situation, without the benefit of a couple of moons'. This was pure fancy on Voltaire's part, as was a similar mention of two Martian satellites in Gulliver's *Voyage to Laputa*, written by Jonathan Swift in 1727.

Several searches for Martian satellites were made, notably by William Herschel in 1783, although these all proved negative. However, the American astronomer Asaph Hall decided to carry out a search for Martian satellites in 1877 at a time when the planet was reasonably close to Earth. Using the 66-centimetre refracting telescope at the U.S. Naval Observatory, located near the Potomac River in Washington, he began his search in early August. On 11 August he caught his first glimpse of a tiny satellite close to Mars on its eastern side. This was Deimos, the outermost of the two moons. Almost as

trial volcanoes by at least 50 times.

The spectacular Valles Marineris is an extensive network of valleys stretching away from the area to the east of Tharsis. Vallis Marineris, named after the Mariner 9 spacecraft which discovered it, straddles some 4,000km (2,485 miles) of the Martian terrain and runs roughly parallel to the equator.

Mariner 9 revealed features that strongly resembled dried-up river beds. While liquid water doesn't exist on Mars today, it may have done in the past. This water is probably locked away under the Martian surface as permafrost, although the northern polar ice cap is known to contain water ice covered by frozen carbon dioxide. During the Martian summer this covering disappears leaving large quantities of water ice behind. Viking images also revealed features resembling areas that have been eroded by huge floods. The water responsible for this may have emerged from beneath the surface following meteoritic impacts. The heat generated by the impact may have melted the ice beneath the surface, the resulting flash floods sending vast torrents of water across the surrounding terrain.

Right: One of the few images of Phobos taken by the Russian Phobos 2 craft before it ceased to function on 27 March 1989.

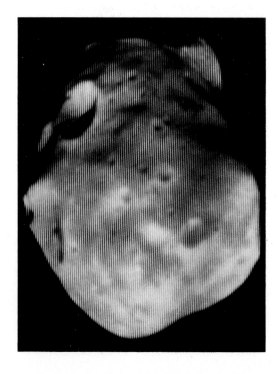

soon as he had noted its position, fog from the nearby river obscured his view.

Bad weather prevented further observation until 16 August when he saw that the object had moved through the sky with Mars and that it was now on the other side of the planet. On the night of 17 August he was checking the position of the satellite when he noticed the innermost satellite Phobos. These observations were confirmed and the news released to the astronomical world on 18 August. The fanciful conjectures of Voltaire and Swift had been proved correct!

Both Phobos and Deimos are tiny, irregularly-shaped worlds, orbiting Mars in circular paths in the plane of the Martian equator. Phobos orbits Mars once every 7 hours and 39 minutes at a distance of 9,380km (5,830 miles). Deimos lies at a distance of 23,460km (14,577 miles) and takes 30 hours 18 minutes to orbit Mars. Both satellites have synchronous rotations, meaning that their periods of axial rotation are equal to their periods around Mars. In other words, they keep the same face turned towards Mars at all times, just as the Moon keeps the same face turned towards Earth throughout its orbit.

Because the orbital period of Phobos is less than the axial rotation period of Mars, tidal forces set up between Mars and the satellite are slowly reducing the distance between the two objects. The result of this is that Phobos is gradually approaching Mars and it is estimated that it will collide with the planet in around 30 million years or so.

Phobos, the larger of the two, measures around 27km × 22km × 19km (16.7 × 13.6 × 11.8 miles) whilst Deimos is just 14km × 12km × 11km (8.7 × 7.5 × 6.8 miles). The sizes of Phobos and Deimos were first approximately determined by the American Mariner 9 probe in 1971–72, which also took the first good pictures of them. Phobos and Deimos were the first satellites of another planet to be photographed from close range.

Both Phobos and Deimos were found to be covered in craters, some of which have been given names. Appropriately enough, the two largest craters on Deimos have been named Swift and Voltaire. The largest crater on Phobos (roughly 10km (6.2 miles) in diameter and nearly

half the major diameter of Phobos itself) is called Stickney (the maiden name of Asaph Hall's wife) in recognition of the encouragement she gave to her husband during his search for the two satellites. Hall himself has had the second largest crater (6 km (3.7 miles) across) named after him.

The surfaces of Phobos and Deimos, although both covered with a layer of debris, or regolith, and both heavily cratered, are very different in other respects. The surface of Phobos plays host to a number of long, parallel grooves up to 200m (650ft) wide and 20m (65ft) deep which probably came into being during the impact event which produced the largest crater and which probably came close to shattering this tiny world.

The surfaces of both satellites are dark grey in colour. They have low albedos, which means that they reflect only a small amount (less than 6 per cent) of the sunlight falling upon them. These two factors make them similar to some asteroids, suggesting that they are indeed asteroids which wandered across the orbital path of Mars and have been captured by the Martian gravity. Their densities are around 2g/cm³, roughly half that of Mars and similar to the densities of some meteorites.

A Viking photograph of Phobos and Deimos.

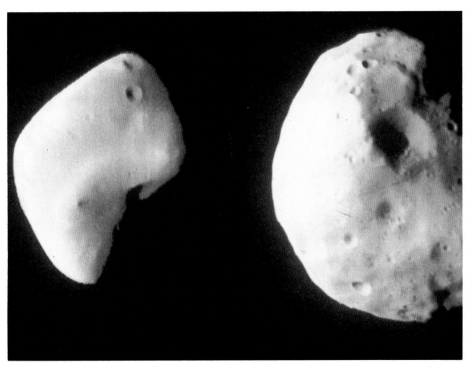

Part II
The Asteroids

The Titius-Bode Law

In 1766 the German mathematician Johann Titius brought to light an interesting numerical relationship linking the distances of the known planets from the Sun. He took the numbers 0, 3, 6, 12, 24, 48 and 96, each of which, apart from 3, has a value twice that of the previous number. He added 4 to each to obtain 4, 7, 10, 16, 28, 52 and 100 following which he divided each of these numbers by 10.

There is nothing remarkable here until comparisons are made with the distances of the planets from the Sun. Taking the Earth's distance as 1, those of the other planets fall in well with the sequence:

Planet	Distance Given by the Law	Actual Distance
Mercury	0.4	0.39
Venus	0.7	0.72
Earth	1.0	1.00
Mars	1.6	1.52
–	2.8	–
Jupiter	5.2	5.20
Saturn	10.0	9.54

Titius remarked that there was no planet to correspond with the value of 2.8 and he suggested that an as yet undiscovered planet was orbiting the Sun between Mars and Jupiter. At first astronomers were sceptical, although in 1772 the German astronomer Johann Bode revived and publicized the idea, so much so that it is now sometimes known, rather unfairly, as Bode's Law, in spite of the fact that credit for its discovery should really go to Titius.

Astronomers still did not take the idea seriously, although many were won over to Bode's way of thinking following the discovery of Uranus by William Herschel in 1781. Uranus was found to tie in well with the Law which gave a distance of 19.6 as compared with the actual value of 19.18.

The Discovery of the Asteroids
The idea of a 'missing planet' now gained in popularity. In 1800 the German astronomer Johann Hieronymus Schroter called together a group of observers at his private observatory in Lilienthal, near Bremen in North Germany. The purpose of this meeting was to organize a

The comparative sizes of the asteroids.

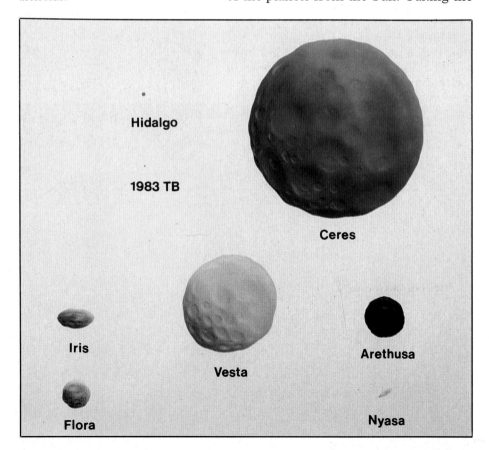

Hidalgo

1983 TB

Ceres

Iris

Arethusa

Vesta

Flora

Nyasa

THE FIRST TEN ASTEROIDS

No.	Name	Discoverer and Place	Date of Discovery	Mean Distance from Sun (au) (*)	Orbital Period (yrs)	Diameter km (miles)	Orbital Inclination (°) (**)
1	Ceres	Piazzi/Palermo	1 January 1801	2.768	4.61	1025 (637)	10.60
2	Pallas	Olbers/Bremen	28 March 1802	2.773	4.61	540 (355.5)	34.80
3	Juno	Harding/Lilienthal	1 September 1804	2.671	4.36	249 (154.7)	13.00
4	Vesta	Olbers/Bremen	29 March 1807	2.362	3.63	555 (345)	7.14
5	Astraea	Hencke/Driesen	8 December 1845	2.577	4.14	116 (72)	5.35
6	Hebe	Hencke/Driesen	1 July 1847	2.424	3.78	206 (128)	14.79
7	Iris	Hind/London	13 August 1847	2.386	3.69	222 (138)	5.50
8	Flora	Hind/London	18 October 1847	2.202	3.27	160 (99.4)	5.89
9	Metis	Graham/Markree	25 April 1848	2.387	3.68	168 (104)	5.59
10	Hygeia	De Gasparis/Naples	12 April 1849	3.138	5.59	443 (275)	3.84

* au = astronomical unit (the mean distance between the Earth and the Sun, equal to 149,597,870km (92,958,348 miles)
** The orbital inclination of any object orbiting the Sun is the tilt (expressed in degrees) of the plane of its orbit with respect to the plane of the Earth's orbit.

THE TEN LARGEST ASTEROIDS

No.	Name	Diameter km (miles)
1	Ceres	1025 (637)
4	Vesta	555 (345)
2	Pallas	540 (335.5)
10	Hygeia	443 (275)
704	Interamnia	338 (210)
511	Davida	335 (208)
65	Cybele	311 (193)
52	Europa	291 (180.8)
451	Patientia	281 (174.6)
31	Euphrosyne	270 (167.7)

An artist's conception of the asteroid Icarus glowing red hot as it approaches the Sun.

search for the planet. Among those present were the Hungarian astronomer Baron Franz Xaver von Zach and the Germans Heinrich Olbers and Karl Harding. The group called themselves the Celestial Police. They decided to approach other astronomers, requesting their help in searching different areas of the Zodiac.

Before the Celestial Police began their search, however, the first of the asteroids was found independently by the Italian astronomer Giuseppe Piazzi on 1st January 1801, from the Palermo Observatory in Sicily. Piazzi was mapping faint stars in Taurus when he noticed what at first appeared to be nothing more than a previously-uncharted star. However, the object was seen to change position over the following few nights and Piazzi realized that he might have stumbled across the missing planet.

Piazzi wrote to von Zach informing him of the discovery although by the time his letter was received the object had moved behind the Sun and was no longer visible. At von Zach's request, however, the German mathematician Carl Friedrich Gauss calculated its orbit from the observations that were available. He published a prediction as to when and where it would reappear and in December, 1801 von Zach spotted it almost exactly where Gauss said it would be. The orbital distance of 412 million km (256 million miles) was found to tie in well with Titius' sequence of numbers, with a value of 2.77. The choice of name for the new planet was left to Piazzi who chose Ceres in honour of the Roman goddess of corn and harvests.

Ceres turned out to be small by planetary standards, and a subsequent search was carried out by the Celestial Police. This resulted in the discovery of three more asteroids. Olbers discovered Pallas in March 1802, while Harding located Juno in September 1804. Olbers found a fourth minor planet, Vesta, in March 1807. After this, no more discoveries were made and the Celestial Police disbanded in 1815.

In 1845, a 15-year search for further asteroids by the German astronomer Karl Hencke came to fruition when he located Astraea. This was followed by his discovery of the sixth asteroid, Hebe, in July 1847 and since then not a year has gone by without further discoveries being made. These early finds were made visually by astronomers who spent many hours at the telescope, under what were often cold and unpleasant conditions, searching carefully for uncharted starlike objects that betrayed themselves by slowly changing their positions from night to night. However, visual searches were about to make way for a far more reliable and successful method.

In 1891 the German astronomer Maximilian Franz Joseph Cornelius Wolf (better known as Max Wolf) made the first photographic discovery of a minor planet. The object in question was Brucia, the 323rd minor planet to be found. This was to be the first of many photographic discoveries. By 1900, the use of astrophotography had brought the total number of known minor planets to 452, Wolf himself going on to discover a total of 232.

By 1923, the list of minor planets had risen to over a thousand. Images of minor planets were found on photographs that were usually part of sometimes completely-unrelated research and astronomers began to consider minor planets as a positive nuisance. One astronomer even went so far as to brand them 'vermin of the skies', a somewhat undeserved title for a class of object that is becoming of increasing interest and importance to astronomers!

Asteroid Formation and Distribution

Ceres, the first asteroid to be discovered, is also the largest, with a diameter of around 1,025km (637 miles). This is equal to almost a third of the total estimated combined mass of all the asteroids. The second largest is Vesta, with a diameter of around 555km (345 miles), closely followed by Pallas, an irregularly-shaped object which is some 540km (335 miles) across. Around a dozen other asteroids have diameters of 250km (155 miles) or more, a further 18 having diameters of over 200km (124 miles).

Even if all the asteroids were collected together, they would form an object less than half the diameter of the Moon.

It was initially thought that the asteroids may be the fragments of a larger object that broke up long ago, although it is clear that, even taking any undiscovered asteroids into account (all of which must be quite small to have escaped detection so far), the original planet must have been tiny. It seems fair to assume that the asteroids are actually debris left over from the formation of the Solar System and which never collected to form a major planet.

Naming Asteroids

During the early years of discovery, mythological names were given to the asteroids, although once the discoveries began to mount up, the supply of names became exhausted. It is interesting to note that, until 433 Eros, all the names given to the asteroids were female! The ranks of asteroid names now covers a wide range of subjects. Some, such as 1125 China, are named after countries while others, including 2001 Einstein and 2039 Payne-Gaposchkin are named after famous people. Stranger are names like that given to 518 Halawe (named after an Arabian sweet) and 1625 NORC (named after the Naval Ordnance Research Calculator at Dahlgren, Virginia). Permanent numbers are assigned to asteroids once their orbits have been calculated and confirmed. Generally speaking, the discoverer of an asteroid has the honour of naming it.

Orbits

Although many thousands of asteroids have been observed, only 3,500 or so have had their orbits accurately determined. It is thought that somewhere in the region of 100,000 asteroids may exist that are bright enough to appear on photographs taken through Earth-based telescopes. Most of those that have been discovered orbit the Sun between Mars and Jupiter, in the region known as the asteroid belt. However, a number have been found outside this region.

Jupiter, with its immense gravitational influence, plays an important role in the distribution of the asteroids. An example of this is seen with the Trojan Asteroids,

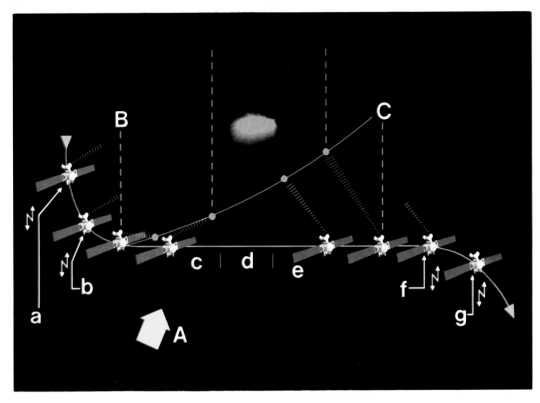

Asteroid Mass Measurement Experiment. In this experiment a test object of known mass is launched in the vicinity of the asteroid; its trajectory is determined and the effects of solar pressure determined (B). As the test mass passes the asteroid, the influence of the latter on the trajectory of the test mass is assessed, allowing the mass of the asteroid to be determined (C). a – Navigation 1; b – Navigation 2; c – Approach; d – Fly-by; e – Departure; f – Navigation; g – Data to Ground Transmission.

two groups of asteroids that travel around the Sun in the same orbit as Jupiter. Each group lies a sixth of the way from Jupiter around its orbit, one group leading and the other trailing behind. The gravitational influences of the Sun and Jupiter act at these so-called Lagrange points to retain asteroids. The Trojan Asteroids are all named after characters in the Trojan War and are all small. Because of their distances, little is known about them.

Asteroids which remain within the main asteroid belt are referred to as Belt Asteroids. However, the distribution of asteroids within the asteroid belt is by no means uniform. In 1857 the American astronomer Daniel Kirkwood suggested that there would be gaps in the asteroid belt, created by gravitational perturbations of Jupiter. The existence of what came to be known as Kirkwood Gaps was confirmed in 1866.

Kirkwood Gaps were created through repeated alignments of asteroids with Jupiter. For example, an asteroid with an orbital period of exactly half of that of Jupiter would, on every second orbit, be aligned with Jupiter. These alignments, together with the angle between the orbital planes of the two objects, would be identical each time. Unlike random perturbations which occur between most of the asteroids and Jupiter, and which tend to cancel each other out over time, these regular perturbations mount up, eventually resulting in the asteroid being pulled away from its original orbital path.

Other asteroids with similar orbits would be affected in the same way and a gap in the asteroid belt would eventually be formed. This gap would be virtually devoid of asteroids. Similar gaps would be formed at other orbital distances where the orbital period of the asteroids would be an exact fraction of that of Jupiter, resulting in similar repeated perturbations. Observations have consistently proved this to be the case.

Other asteroid groups include the Amor Asteroids and the Apollo Asteroids. Each group is named after the first of its type to be identified. Both groups have elongated orbits which take them across the orbital paths of major planets. The Amor Asteroids cross the orbit of Mars while the Apollo Asteroids cross that of Earth.

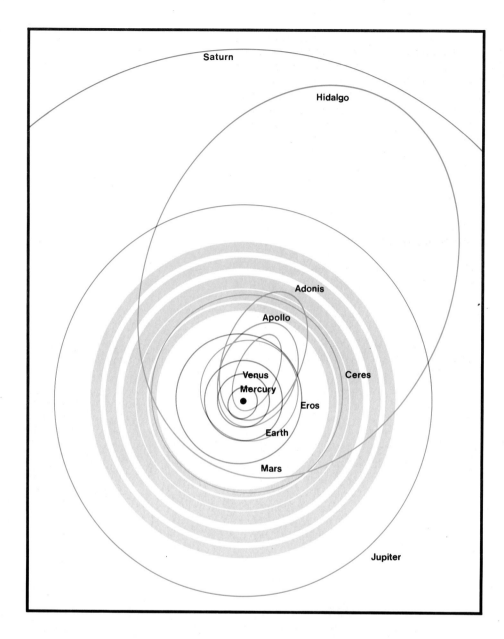

Apollo Asteroids are occasionally seen to pass close to the Earth. Notable examples of close-approach asteroids are Hermes and Eros. In October, 1937 Hermes came within 900,000km (560,000 miles) of our planet (a little over twice the distance of the Moon) while Eros makes regular close approaches to Earth, the closest it can come being around 23 million km (14.3 million miles). The most recent close approaches of Eros took place in 1931 and 1975. Hermes has since been lost, although the orbital path of Eros is well known. Close as Hermes was, the record for an asteroid 'near miss' is held by object 1989FC which came within 700,000 km (435,000 miles, less than twice the Moon's distance) on the night of 22/23 March 1989. The asteroid, thought to be several hundred metres in diameter, was not actually noticed until short trails appeared on photographs taken on 31 March, these forming part of a search for near-Earth asteroids. When the path of 1989FC was plotted back it was realized how close to Earth it had come. An orbit for 1989FC has been calculated. It orbits the sun every 1.03 years, roughly between the orbital distances of Venus and Mars. If an impact had occurred a crater several kilometres in diameter would have been formed and a great deal of damage and loss of life could have resulted had the object landed in a populated area. It is estimated that the speed of impact would have been just over 15km/sec (9.3 miles/sec).

The regular close approaches of Eros and other close-approach asteroids, enabled astronomers to examine them in more detail. Rates of axial rotation can be

Above: The orbits of the asteroids. Most of the asteroids move in the region between the orbits of Mars and Jupiter. Hidalgo has a highly inclined and eccentric path much like a comet. Apollo is known as an Earth-grazer.

Right: The asteroid Eros taken on 21 January 1975 and seen near Pollux, the bright star at the centre.

calculated by measuring variations in its brightness. As an asteroid rotates, different types of surface terrain are presented to us. Observation of a regular cycle of variations helps to determine the rotation period. Eros itself spins once every 5.27 hours.

Many asteroids orbit the Sun in groups, or families, the members of which are thought to have originated from collisions between larger asteroids. The speed at which the objects meet determines the outcome of the collision. For example, high speed impacts may produce the many fragments typical of an asteroid family. However, if the collision speed is low, the chunks may actually collect together again under the influence of gravity. Intermediate between these two scenarios is the formation of binary asteroids, with two components orbiting each other. A number of binary asteroids are known to exist, the first to be discovered being 532 Herculina. Observation of the occultation of a star by Herculina in June 1978 revealed a second occultation immediately after the first. This second event was due to a satellite orbiting Herculina. The diameters of the two objects are around 215km (133.5 miles) (Herculina) and 50km (31 miles) (the satellite), the diameters being calculated from observation of the occulation.

Some asteroid collisions throw off fragments, known as meteoroids, which eventually land on other planetary members of the Solar System. Impact craters are formed by these objects (see 'Meteors and Meteorites'), the size of which depends upon the size of meteoroid and the speed of impact. Thousands of impact craters have been observed throughout the Solar System, ranging from the surface of Mercury out to the satellites of the remote planets. The most familiar are those scattered across the lunar highlands.

The Arizona Meteor Crater, created by the impact of an asteroid fragment.

Part III
The Outer Planets

The Voyager Missions

A Titan-Centaur vehicle lifts the Voyager 2 spacecraft towards its rendezvous with the outer planets, 20 August 1977.

The exploration of the outer gas giants of our Solar System has been designed in three stages. An initial reconnaissance of Jupiter and Saturn was performed by Pioneer 10 (Jupiter) and Pioneer 11 (Jupiter and Saturn), launched respectively in March 1972 and April 1973.

Stage Two, the Voyager missions, were made possible through a favourable alignment of the outer planets that occurs only once every 179 years. As a result, Voyager 2 was able to visit Jupiter, Saturn, Uranus and Neptune all within a 12-year period by utilizing what is called the 'gravitational sling shot' effect. During each planetary encounter, the gravitational field of the planet was used to accelerate the craft towards its next objective. The alignment of the planets meant that each of the next planets would lie directly in front of the probe as the latter approached its orbit.

Stage Three will be individual missions to Jupiter and Saturn. Jupiter will be visited by the Galileo probe, while Saturn and its moon Titan will be explored in greater detail by the Cassini mission (see 'Future Exploration of the Solar System').

The Voyager Craft

The Voyager craft are the most sophisticated automatic space probes ever built. Weighing almost a tonne, they each carry eleven instruments including wide-field and narrow-angle cameras and antennae for measurements at radio wavelengths. Because of the immense distances covered by the craft, there was a delay between the transmission and receipt of

signals between the probes and Mission Control, almost three hours in the case of the Uranus encounter in spite of the fact that the signals were travelling at the speed of light, 299,792 km/sec (186,282 miles per second). Commands had to be sent 'in advance' of the times they were to be implemented, and it was impossible to direct the missions in real time. To overcome this, each craft carried powerful, partially-automatic computers. Their roles included automatic stabilization of the vehicle and the storage and implementation of command sequences sent previously from Earth.

The immense distances covered by the Voyager probes meant that power for the on-board computer and instruments had to be generated on the craft. Unlike probes sent to the terrestrial planets, which use solar panels to collect energy from the Sun for conversion to electrical power, the Voyager craft were to venture into the outer regions of the Solar System. From there, the Sun's energy would be too weak for solar panels to be effective. Each of the craft is equipped with Radioisotope Thermoelectric Generators (RTGs) which use the heat released by the decay of the radioactive isotope Plutonium 238 to create the 450 watts of power needed by the Voyager instruments and sub-systems. The RTGs are mounted on booms to keep them well away from the instruments, thereby minimizing any possibly harmful radiation effects. It is expected that sufficient power will be available until well into the 21st century.

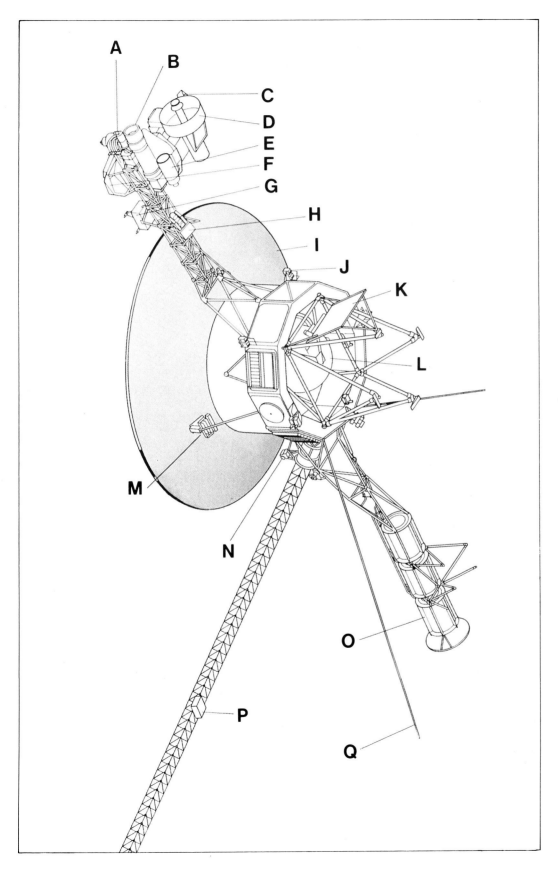

Eleven experiments could be carried out by remote-controlled instruments on Voyager 2. A – Wide-angled camera; B – Narrow-angle camera; C – Ultraviolet spectrometer; D – Infra-red spectrometer; E – Photopolarimeter; F – Plasma detector; G – Cosmic ray detector; H – Low-energy particle detector; I – High-gain antenna; J – Thruster; K – Optical calibration target; L – Fuel tank; M – Sun sensor; N – High-field magnetometer; O – Radioisotope-thermoelectric generators; P – Low-field magnetometer; Q – Radio-astronomy and plasma-wave antenna.

Voyager's original flight plan called for encounters with Jupiter and then Saturn. While Voyager 1 has headed upwards and outward after its Saturn encounter in late 1980, Voyager 2 has since reached Uranus and Neptune.

Mission Highlights

Voyagers 1 and 2 were launched in 1977 from Cape Kennedy, Voyager 2 on 20 August and Voyager 1 on 5 September. Voyager 1 overtook its companion en route to Jupiter while crossing the asteroid belt. The Voyager 1 craft was subject to teething troubles during the flight to Jupiter, including the temporary jamming of its revolving platform. Encounter with Jupiter took place in March 1979 during which excellent images of the planet were returned together with our first detailed look at the moons Io, Ganymede and Callisto, all of which were revealed as individual and fascinating worlds. The Jovian ring system was discovered by Voyager 1. The Saturn encounter followed in November 1980, during which images of Titan, Rhea and Mimas were obtained.

Though Voyager 1 was virtually trouble-free, Voyager 2 was not so lucky. The original spacecraft had to be replaced prior to launch (three had actually been constructed) following which there were problems both during the actual launch and the early part of its flight. In April 1978 the primary radio receiver failed totally and the secondary radio receiver suffered technical problems, resulting in the craft being unable to track the commands sent from Earth. NASA even-

tually re-established contact with Voyager 2, following which it was able to carry out the most exciting mission ever carried out by a space probe.

Once the Voyager 2 craft reached Jupiter it was used to follow up the results of Voyager 1. Closer views of the volcanoes on Io and the ring system were obtained as well as high resolution pictures of Europa, the Galilean satellite least observed by Voyager 1. However, there were difficulties encountered at Saturn after Voyager 2 had crossed the ring plane. The scan platform, carrying most of the instrumentation, locked in position and much of the data obtained during the period following encounter was lost, including high resolution images of Tethys and other satellites. The problem was resolved after 3 days.

Following the encounter with Saturn, it was decided to send Voyager 2 on towards an encounter with Uranus, which took place in January 1986. This was a remarkable achievement considering that the craft had undergone a nine year flight over an immense gulf of space and that Uranus was a 'moving target' with a diameter of only just over 50,000km (31,069 miles).

The success enjoyed during the Jupiter and Saturn encounters was continued at Uranus. Detailed images of both the planet and its ring system were obtained, together with the discovery of additional rings and a further 10 satellites orbiting the planet.

The Voyager 2 craft was then re-targeted to Neptune, which it encountered in August 1989. All our former knowledge of this distant world had been obtained from Earth-based observation and was scanty to say the least. We knew of two satellites in orbit around the planet, with a third suspected. This number was raised to eight by Voyager 2. Also, the presence of a ring system around Neptune was confirmed.

The Deep Space Network

In order to track the Voyager spacecraft, and indeed other interplanetary probes, NASA uses a system of radio receivers scattered across the globe. Known as the Deep Space Network (DSN), it employs nine deep space communications stations on three continents, these being Australia (Canberra), Europe (Madrid) and the U.S.A. (Goldstone, California). The three main stations are widely separated to ensure that the spacecraft being monitored remains 'in view' of at least one station. Communication and data links between the ground stations are

The 210-ft dish antenna at Goldstone, California, used as part of the Deep Space Network.

provided both by Earth-orbiting satellites and ground-based networks. The signals received by each individual antenna are combined into a single, stronger signal by a process known as arraying.

As probes travel farther out into space, the strength of the signals transmitted back to Earth weakens. The signals received from Voyager 2 at Neptune, around 4,500 million km (2,796 million miles) away, were roughly half as strong as those received from Uranus, at a distance of about 3,000 million km (1,864 million miles). For the Voyager 2 encounter with Uranus in January 1986 the Canberra complex was arrayed with the Australian government's Parkes Radio Observatory 275km (171 miles) away, while for the encounter with Neptune in August 1989, improvements were made to the DSN itself. These included the upgrading of the three 64-metre diameter antennae to 70 metres, enabling them to detect the fainter transmissions more effectively, and the construction of an additional 34-metre antenna at the Madrid complex. Together with these improvements, the Very Large Array (VLA) was used to track Voyager 2. The VLA consists of 27 identical 25-metre antennae, the signals received by which can be electronically combined (arrayed) to effectively form a single large radio telescope. The antennae are mounted on railroad tracks in a Y-shape, spread out across the plains of St Augustin, around 80km (50 miles) west of Socorro, New Mexico. The antennae can be moved around to produce configurations best suited to the type of observation being carried out. The signals received by the VLA were transmitted via satellite to Goldstone where they were combined with those received at Goldstone itself.

Towards the Stars

Their planetary encounters accomplished, both Voyager craft are currently on their way out of the solar system. One of their objectives is to identify the location of the heliopause, the region at which the solar wind and interplanetary magnetic field gives way to true interstellar space. Deviations in the trajectories of the Voyager probes (and the Pioneer 10 and 11 probes which are also heading out of the Solar System) may also help astronomers track down the position of the planet thought by some to be orbiting the Sun beyond Pluto and called Planet X.

The chances of the Voyager probes encountering another planetary system, and being captured by an extraterrestrial civilization, are extremely remote. However, in the event of this happening, each probe carries a videodisc together with instructions for operation. These discs contain information about the Earth and its inhabitants. It will be many centuries, however, before either probe passes close to another star.

An artist's conception of the European Southern Observatory.

Jupiter

With a diameter of over 11 times that of the Earth, Jupiter is the largest planetary member of the solar system. Its volume exceeds that of the Earth by more than 1,400 times and this giant planet contains around 70 per cent of the entire planetary mass of the Solar System. Jupiter travels around the Sun once every 11.86 years at a mean distance of just over 778 million km (483 million miles).

Jupiter not only has an axial rotation shorter than any of the other planets but it also exhibits differential rotation. Equatorial rotation is 9h 50m 30s as compared to 9h 55m 41s in the polar regions. This was first observed for Jupiter by the French astronomer Giovanni Domenico Cassini in 1690.

This short rotation period has resulted in Jupiter becoming noticeably flattened, or oblate. The equatorial diameter is 143,800km (89,350 miles), 8,000km (4,971 miles) greater than its polar diameter. As Jupiter spins on its axis, each part of the planet undergoes an outward-acting centrifugal force, this force being greater the further away a region is from the axis of rotation. The outlying equatorial regions have been forced outwards far more than have the polar regions, which are closer to the axis of rotation. Jupiter's pronounced equatorial 'bulge' is visible through most telescopes.

Inside Jupiter

There are three main regions to the Jovian interior. At the centre of the planet is a rocky core comprised of iron and silicates which in turn is surrounded by a layer of liquid metallic hydrogen some 40,000km (24,855 miles) deep. Within this region, the immense pressure exerted by the overlying Jovian atmosphere has resulted in the hydrogen atoms having been stripped of their single electrons. The Jovian atmosphere has a 20,000km (12,427 miles) deep envelope of hydrogen, the outer layer of which is around 1,000km (621 miles) deep. The colourful Jovian cloudtops visible through telescopes are situated on top of this outer layer.

Jupiter is one of the brightest planets and is a colourful sight, its disc being

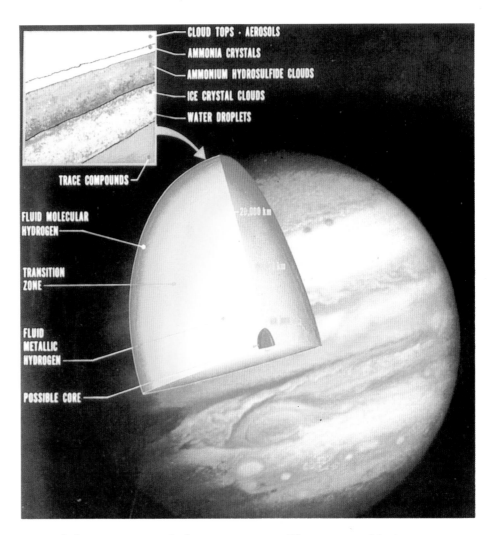

CLOUD TOPS · AEROSOLS
AMMONIA CRYSTALS
AMMONIUM HYDROSULFIDE CLOUDS
ICE CRYSTAL CLOUDS
WATER DROPLETS
TRACE COMPOUNDS
FLUID MOLECULAR HYDROGEN
TRANSITION ZONE
FLUID METALLIC HYDROGEN
POSSIBLE CORE

The structure of Jupiter.

crossed by numerous belts or zones which exhibit a multitude of colours including red, orange, brown and yellow. The brighter zones are regions where gases from within the planet are rising to the surface to cool, while the darker belts are regions where material is descending.

Moderate telescopes will show a number of these zones, together with the Great Red Spot, a huge atmospheric feature located in the Jovian southern hemisphere and observed almost constantly since it was first seen by Cassini in 1665. The size of the Great Red Spot has been seen to vary over time. At its maximum extent it can attain a length of 40,000km (24,855 miles), over three times the diameter of the Earth! The maximum width of the Great Red Spot is around 14,000km (8,700 miles). Its exact cause has yet to be confirmed but the Pioneer and Voyager missions have

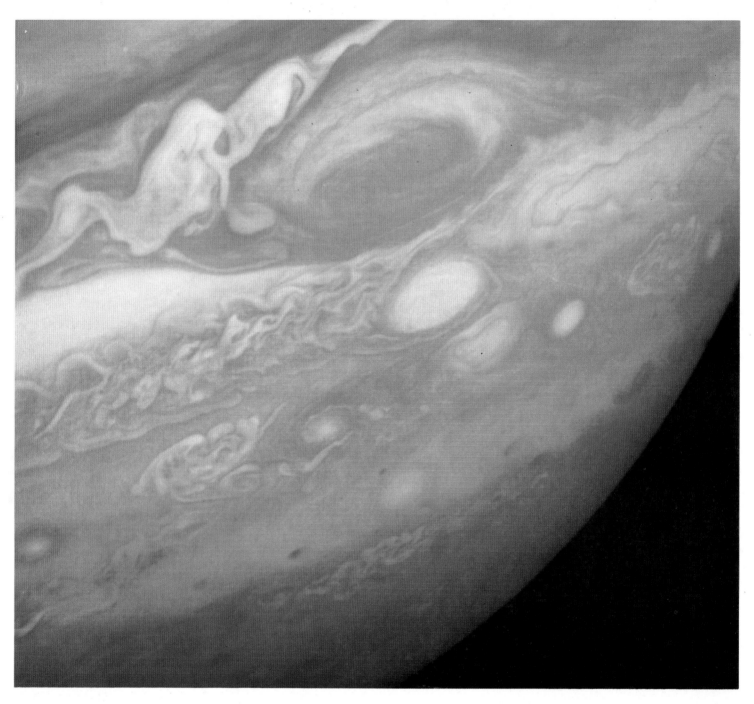

View of Jupiter's Great Red Spot taken by Voyager 1 in February 1979. The colourful wavy patterns below and·to the left of the Spot are believed to be regions of complex atmospheric wave motion.

provided results which suggests that it is a whirling storm whose red colour may be caused by the presence of red phosphorus. The Galileo mission to the planet (see 'The Future of Planetary Exploration') may tell us more about this remarkable feature.

The Pioneer and Voyager probes sent back beautiful images of the Jovian atmosphere, although the earlier Pioneer photographs were of an inferior standard to the later Voyager pictures. However, it was clear that, during the intervening period, changes had taken place in Jupiter's atmosphere. The most notable of these was in the region around the Great Red Spot. Throughout the past 300 years, Earth-based astronomers have witnessed changes in the size and colour of this feature, although more subtle

changes were apparent after closer examination by Pioneer and Voyager.

At the time of the Pioneer 10 and 11 flybys (in December 1973 and December 1974), the Great Red Spot was surrounded by a planet-girdling white zone which occupied a significant area of the Jovian southern hemisphere. The situation had changed dramatically by 1979 when Voyager 1 (March) and Voyager 2 (July) flew past the planet. The Great Red Spot had been crossed by a dark belt, and there was an increase in turbulence around the area. Measurements revealed that the Great Red Spot rotates anticlockwise over a period of around six days. The winds to the north and south of the Spot blow in opposite directions, seemingly fuelling the Great Red Spot's rotation.

A close-up view of the cloud motions seen in the previous picture. The colours of the clouds are related to their chemical composition and temperature.

Four of the Jovian satellites. From top left, Callisto, Ganymede, Europa, Io.

The Satellites of Jupiter

The four largest Jovian satellites were discovered in 1610 by the Italian astronomer Galileo Galilei. Known collectively as the Galilean satellites, Io, Europa, Ganymede and Callisto orbit Jupiter in virtually circular paths almost exactly above Jupiter's equator. Although they were discovered almost four centuries ago, little was known about the Galilean satellites until the Jovian system was visited by the Voyager 1 craft in March 1979. Prior to this, we knew nothing about their surfaces, our knowledge extending only to their spatial characteristics (diameters, orbital periods, distances from Jupiter and so on). However, the cameras of Voyager 1 (and Voyager 2 which reached Jupiter four months later) have revealed a wide diversity of surface features on these worlds.

A major surprise for the Voyager scientists was the discovery that Io was a volcanically-active world. Voyager 1 returned pictures that showed eight volcanic eruptions in progress with volcanic plumes extending to heights of up to 280km (174 miles). The surface of Io was found to be very young with no impact craters being detected. However, volcanic centres were widespread, some calderas having diameters of up to several tens of kilometres.

Lava flows were seen extending from the calderas for distances of up to several hundred kilometres. It is this perpetual volcanic activity which continually deposits sulphur compounds onto the surface from inside Io, thereby masking any other type of geological feature. Little surprise, therefore, that no impact craters were observed by Voyager.

Europa is a highly reflective world and was found to be covered by a layer of water ice estimated to be around 100km (62 miles) thick. Very few impact craters were seen which indicates either that they never existed or that they have been obliterated by water ice rising from below the surface. Huge dark linear features up to 70km (43.5 miles) wide extend for hundreds, in some cases thousands, of kilometres across the surface of Europa, giving Europa the appearance of a fractured world. However, these features may not all be fractures, as some brighter lines appear to rise above the surrounding terrain.

Crusts of water ice were also found on Ganymede and Callisto although, unlike Io and Europa, large numbers of craters were present. With a diameter even greater than Mercury, Ganymede is the largest Jovian satellite and the largest satellite in the Solar System. Its surface is composed of light and dark terrain, the dark areas containing a larger number of meteorite impact craters. Together with its craters, Ganymede has regions of grooved terrain, which may have formed from material breaking through the crust. The largest feature seen on Ganymede is Galileo Regio, a dark circular area of ancient crust 4,000km (2485

This unique photograph was taken by Voyager 1 on 4 March 1979, and shows an enormous volcanic explosion occurring on Io. It is estimated that solid material was thrown to a height of 100 miles.

Physical Data For the Galilean Satellites

	Diameter km (miles)	Orbital period (days)	Mean distance from Jupiter km (miles)
Io	3,630 (2,255)	1.77	421,600 (261,977)
Europa	3,138 (1,950)	3.55	670,900 (416,889)
Ganymede	5,262 (3,270)	7.16	1,070,000 (664,885)
Callisto	4,800 (2,982)	16.69	1,880,000 (1,168,209)

Photograph of the southern hemisphere of Jupiter obtained by Voyager 2 on 25 June 1979. Seen in front of the turbulent clouds is Io, the innermost of the large Galilean satellites of Jupiter.

miles) in diameter which, like other similar but smaller areas, contains an abundance of craters.

Callisto is the darkest of the Galilean satellites and its surface shows large numbers of impact craters, making it one of the most cratered objects in the Solar System. By far the largest feature on Callisto is the Valhalla Basin, a huge formation probably caused by the impact of an asteroid-sized object. The Valhalla Basin is fairly flat and contains a series of concentric rings 50–200km (31–124 miles) apart and extending out to a radius of around 1,500km (932 miles) from the point of impact. Valhalla is covered by numerous, smaller craters, which suggests an age in the region of 4,000 million years.

Left: Callisto, outermost of the Galilean satellites. It is the least reflective of the four as well as being the most heavily cratered.

The Lesser Satellites of Jupiter

Jupiter has a total of 16 satellites, although others may await detection. As we have seen, the Galilean satellites orbit Jupiter very near the plane of its equator. Another group, all considerably smaller than the Galilean satellites, orbit the planet inside the path of Io.

This latter group, comprised of Metis, Adrastea, Amalthea and Thebe, are the innermost of the Jovian satellites. The largest of these is Amalthea, which was discovered in 1892 by the American astronomer Edward Emerson Barnard. Amalthea was discovered by visual observation through a telescope, but all subsequent discoveries have been made either photographically or by space probe. Until the Voyager probes passed Jupiter in 1979, Amalthea was thought to be the innermost satellite, but Metis and Adrastea have been found to be closer still to Jupiter.

Amalthea is a dark, heavily-cratered and irregularly-shaped world with a strong reddish colour. It is difficult to see at all from Earth, and its discovery was a notable achievement. Metis and Adrastea orbit Jupiter quite close to the Jovian rings and, although only small, their gravitational influence doubtless has

important effects on the particles forming the ring system, probably limiting the outer extent of the rings.

The other eight Jovian satellites orbit the planet in two distinct groups. As with the innermost group, they are all tiny

Above: This photograph provided the first evidence of a ring around Jupiter (Voyager 1, 4 March 1979).

objects, and are probably minor planets that were captured by Jupiter's gravity. The innermost satellites probably formed through the accretion (building up) of smaller particles in the Jovian environment.

Unlike the innermost eight satellites, all of which move around Jupiter virtually in the planet's equatorial plane, the orbital planes of Leda, Himalia, Lysithea and Elara are inclined at between 24 and 29 degrees to the Jovian equator. Their orbits range between 11 and 12 million km (6.8–7.4 million miles) from Jupiter. The orbits of these four satellites are a little eccentric. In other words, unlike the almost-circular orbits of the eight innermost satellites which enable them to keep at almost the same distance from the Jovian cloudtops, the distances from Jupiter of Leda, Himalia, Lysithea and Elara vary between 10 and 20 per cent.

Further out still are Ananke, Carme, Pasiphae and Sinope. These four satellites orbit Jupiter at average distances of between 21 and 24 million km (13–15 million miles) in highly eccentric orbits. Unlike the Galilean satellites, little is known about the smallest members of Jupiter's satellite family, and it remains for future missions to the planet to tell us more.

The Rings of Jupiter
On 4 March, 1979 the Voyager 1 cameras detected a ring system around Jupiter. The discovery of a Jovian ring system, following as it did the discovery of rings around Uranus in 1977, led astronomers to believe that there may be ring systems around all the gas giant planets rather than just around Saturn.

Unlike the rings of Saturn, the Jovian ring system is fairly dark, extremely faint and invisible from Earth. The system is thought to be only a billionth the density of that of Saturn, and it is little wonder that the discovery had to wait until Jupiter was visited by the Voyager 1 space probe.

A few years prior to the visit of Voyager 1, instruments on board the visiting Pioneer 11 spacecraft had detected irregularities in the number of charged particles moving around Jupiter at different distances from the planet. A number of possible explanations for this

were put forward, one of these being that Jupiter had a ring system, the components of which orbited Jupiter in the regions containing the fewest particles.

The Voyager 1 cameras, coupled with observations made by Voyager 2 during its flight through the Jovian environment a few months later, have provided us with the little knowledge we have of the Jovian rings. Following the Voyager 1 discovery, the Voyager 2 craft crossed the plane of the Jovian rings twice, once before and once after its encounter with the planet, and a number of pictures were obtained. What the rings are made of, however, and how large or small the particles are remains to be discovered. What is known is that the system is composed of two main sections. The innermost of these is around 5,000km (3,107 miles) wide and has a brighter, 800km (497 miles) ring outside it. They are around one kilometre thick and extend from between 50,000km (31,070 miles) and 53,500km (33,244 miles) above the outer visible surface of Jupiter. From within the inner ring a region of much thinner material reaches down to the Jovian cloud tops. The entire ring system is enclosed in a tenuous envelope, or halo.

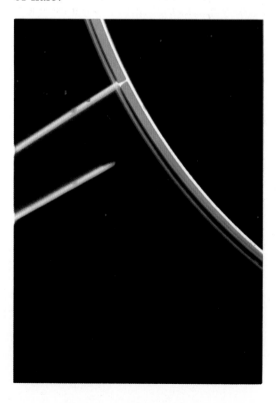

Jupiter's faint ring system is shown in this composite photo as two orange lines at the left. The lower part of the ring is cut short because of Jupiter's shadow.

Saturn

Saturn is visible to the naked eye as a yellowish, starlike point of light. It is quite conspicuous and even a small telescope will reveal Saturn's beautiful ring system and its largest satellite, Titan. To many observers, their first telescopic view of the planet Saturn is an event they will never forget. Although many pictures of this celestial showpiece have appeared, there is nothing to compare with a first-hand sight of the planet, set against the backdrop of a velvet-black sky.

Saturn is the second largest planet of the Solar System, with an equatorial diameter of 120,660km (74,980 miles). Its rapid equatorial rotation period of 10h 13m 59s has resulted in the planet being noticeably oblate, with a polar diameter of just 108,600km (67,480 miles). This degree of polar flattening exceeds even that of Jupiter.

Saturn has a total mass of over 95 times that of the Earth and, with the exception of Jupiter, outweighs all the remaining planets put together. However, one peculiarity of Saturn is its low density, amounting to just under 0.7gm/cm³, roughly half that of Jupiter. This low density means that Saturn must be com-

posed of light elements, and that the planet would actually float in water! Saturn orbits the Sun once every 29.46 years at a mean distance of 1,427 million km (886.7 million miles).

Above: Saturn as Voyager approached it at 18 million km (11 million miles).

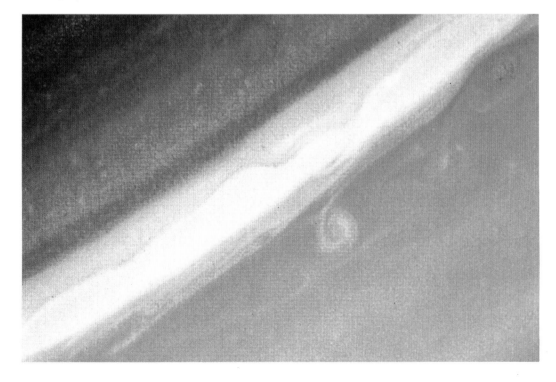

Image of Saturn's northern mid-latitudes. A strangely-curled cloud attached by a thin 'ribbon' to the white band above it was monitored for seven rotations around the planet.

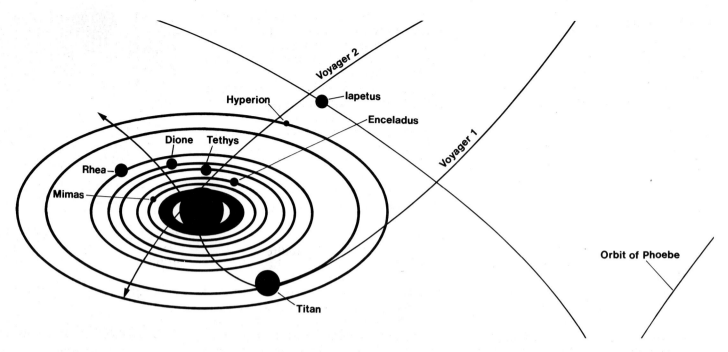

Above: Saturn has 17 known moons and the diagram shows their orbits. All but two of them, Iapetus and Phoebe, lie in the equatorial plane of the planet. Phoebe's rotation about Saturn is clockwise, all the others go counter clockwise.

Right: Voyager 2 photo of Titan, the largest satellite of Saturn. It has a reddish, cloudy atmosphere containing methane and possibly ammonia and hydrogen.

Inside Saturn

Like Jupiter, Saturn has an outer layer of hydrogen, although this is thought to be somewhat thicker than that of Jupiter. Beneath this outer layer is a region of liquid metallic hydrogen which in turn envelopes a rocky core. This core is believed to be much larger than Jupiter's, based on observations that although Saturn has a longer axial rotation period, its degree of oblateness is greater than that of Jupiter. This can only mean that there is a greater concentration of mass at the centre of the planet. Saturn's core contains around 26 per cent of the total mass of the planet, as opposed to around 4 per cent for Jupiter.

As with Jupiter, the outer visible layer of Saturn's atmosphere is crossed by dark belts and bright zones, although these are nowhere near as conspicuous or colourful as those of Jupiter. Gravity plays an important part here. The vastly greater pull of Jupiter's gravity has compressed the outer cloud layers into a total depth of around 75km (45 miles). As a result, the layers are more visible than those of Saturn. Here the gravitational pull is much weaker, resulting in a combined depth of the layers of around 300km (190 miles).

The composition of the atmosphere is also different, that of Jupiter containing 82 per cent hydrogen and 17 per cent helium compared to Saturn's 88 per cent hydrogen and 11 per cent helium. In each case, the atmospheric content includes 1 per cent of other elements. It has been suggested that the lack of helium is caused by Saturn cooling down more quickly than Jupiter. The effect of this cooling was to produce helium 'rain' which fell from the atmosphere towards the inner regions of the planet.

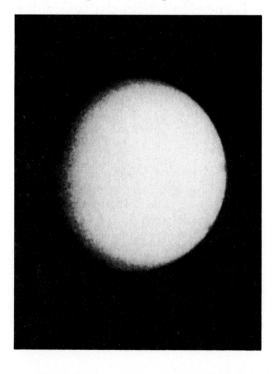

The Saturnian Satellites

Titan

Of Saturn's 21 known satellites, the largest is Titan with a diameter of 5,150km (3,200 miles), which was discovered by the Dutch astronomer Christiaan Huygens in 1655. Titan orbits Saturn at a mean distance of 1,222,000km (760,000 miles). It had long been thought that Titan has an atmosphere, a fact which was confirmed by Gerard Kuiper in 1944 when he detected methane around the satellite. Measurements taken by Voyager suggest that nitrogen (90 per cent) is the most abundant gas in Titan's atmosphere, followed by methane and argon. The atmosphere is so thick that the Voyager cameras were unable to see Titan's surface.

The surface pressure on Titan is 60 per cent greater than that on Earth. Combine this with the surface temperature of −178°C, and conditions are near the triple point of methane, this being the temperature and pressure at which a substance could exist as either a gas, liquid or solid. In the case of the Earth, the temperature and pressure are near the triple point of water which, as we know, exists as a solid (ice), liquid (water) and gas (steam).

Titan's polar regions may therefore be composed of methane ice, with rivers and lakes of methane existing at lower latitudes. There may also be falls of methane rain or snow. Although there are speculations at this time, the Cassini mission to Saturn (see 'The Future of Planetary Exploration'), may provide verification. The mission includes a probe which will descend through Titan's atmosphere sending back results both during the descent and, hopefully, from the surface.

★   ★   ★   ★

Prior to the arrival of Pioneer 11 (September 1979), Voyager 1 (November 1980) and Voyager 2 (August 1981) at Saturn, only ten satellites were known to exist around the planet. The results of these missions combined with improved Earth-based telescopic observation have increased this number. Our knowledge of the known satellites has also substantially improved following the probe missions since many of them have now been studied with a resolution equal to the best Earth-based telescopic observations of the Moon. Apart from Titan, there are a further six satellites with substantial diameters, these being Mimas (392km/ 243 miles), Enceladus (500km/310 miles), Tethys (1,060/658 miles), Dione (1,120km/696 miles), Rhea (1,530km/ 950 miles), Iapetus (1,460km/907 miles) and Phoebe (220km/136 miles).

Mimas

Mimas is the closest satellite to Saturn, orbiting the planet in less than 23 hours. This close proximity makes Mimas very difficult to observe from Earth, although the Voyager cameras revealed a number of features on this satellite. Particularly prominent is a huge crater, named Herschel, which has a diameter a full one-third (130km/80.7 miles) that of Mimas itself! It is thought that Herschel was formed by an impacting object around 10km (6.2 miles) in diameter. Had this object been any larger then the satellite might well have been broken up. As it is, the surface of Mimas contains fractures of the impact, together with many other craters.

Enceladus

Enceladus was shown by Voyager 2 to have quite a young surface, the craters that were formed during the early bombardment of the satellite having been covered by subsequent geological activity. Several areas of the surface were virtually devoid of craters. The surface of the satellite was also seen to contain a number of different types of formation including ice flows, faults and striations, the latter probably being due to the crust of Enceladus being thin and lying on top of a molten interior. This molten interior must come about through internal heat. The source of this internal energy remains something of a mystery.

Tethys

Tethys, like Mimas, has a surface covered with craters, the density of cratering on the two worlds being roughly the same. As with Mimas, there

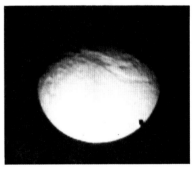

Saturnian satellites. From the top, Dione, Mimas, Rhea (colour enhanced) and Tethys.

A Voyager 2 mosaic of the Saturnian moon Enceladus made from images taken through clear, violet and green filters. Some regions of Enceladus show impact craters up to 35km (20 miles) in diameter.

different types of geological feature including faults, valleys and depressions. The distribution of craters over the surface of Dione varies, due, no doubt, to the covering of different areas by geological activity which has taken place since the meteoritic bombardment early in the history of the solar system. The largest craters are around 100km (62 miles) across, and bright streaks that are seen radiating from some craters are the result of material being ejected after impact.

Rhea, Iapetus and Phoebe

Rhea is second only in size to Titan and its surface plays host to many craters. The surface of Rhea is very old and has not been changed by geological activity as is the case with Enceladus. The craters that were imaged by the Voyager cameras appear to be virtually unchanged since their formation.

Iapetus and Phoebe are the two outermost of the Saturnian satellites. Iapetus orbits Saturn at a mean distance of over 3.5 million km (2.17 million miles) while Phoebe is over three times as far away again. Iapetus was discovered by the Italian astronomer Giovanni Domenico Cassini towards the end of the seventeenth century. Cassini noticed that Iapetus always appeared darker when positioned at one side of Saturn than the other. This was explained by assuming that Iapetus had dark and bright hemispheres, a fact that was indeed shown to be the case by Voyager. One side of Iapetus is bright and appears to be covered with ice or similar material. This brighter hemisphere also has many craters unlike the darker side. This latter side is coated with a dark material that may have originated from within the satellite, a fact suggested by some of the craters seen on the brighter hemisphere having floors containing some of this dark material. Certainly Iapetus is a puzzle for astronomers!

Phoebe, the outermost Saturnian satellite, takes 550 days to travel once around the planet. Phoebe is spherical and has a rather dark, reddish surface which reflects only a small percentage of the sunlight falling on it. Its orbit is retrograde, which has led astronomers to the conclusion that this tiny satellite may once have been an asteroid that wan-

is one very prominent crater. Named Odysseus, this huge feature has a diameter of 400km (248.5 miles), making it larger than Mimas itself! However, unlike the crater Herschel on Mimas, which is some 10km (6.2 miles) deep, Odysseus is quite shallow, its floor having been covered by ice flows. A huge valley was detected on Tethys. Ithaca Chasma has a length in excess of 2,000km (1242 miles) and stretches some three-quarters of the way around the surface of Tethys. In places, Ithaca Chasma is 100km (62 miles) wide and is reminiscent of the Martian Valles Marineris. However, the origin of Ithaca Chasma isn't known with certainty, although it could have been formed as a result of the impact which created Odysseus.

Dione

Dione is similar in size to Tethys and, like Enceladus, displays a number of dered too close to Saturn and was pulled into orbit around the planet.

★   ★   ★   ★

Most of Saturn's 14 other satellites are irregularly-shaped and, although many of them were imaged by Voyager, their diameters are not yet known with certainty. Three of the smallest satellites (Atlas, Prometheus and Pandora) were discovered in 1980 from Voyager photographs. As with Jupiter, the retinue of satellites around Saturn resembles a miniature Solar System, and future exploration promises to reveal many more fascinating secrets about the planet and the satellites orbiting it.

The Rings of Saturn

When the Italian astronomer Galileo Galilei first turned his telescope towards Saturn in 1609–10 he saw little detail on the planet. However, he did observe that the planet appeared to have two lumps, one on either side of the planetary disc. He recorded what he had seen, although their identity remained a mystery.

As time went on telescopes improved, and in 1655 Christiaan Huygens put forward the theory that what Galileo had seen was a system of rings girdling the planet but which, through Galileo's inferior telescopes, appeared as he had described them. Subsequent observation confirmed Huygens' theory.

From Earth we can see the main sections of the ring system, including the bright Ring B, the fainter Ring A and the dim Crepe Ring, otherwise known as Ring C. When the ring system is suitably inclined towards us, the famous Cassini Division comes into view. This dark division separates Ring A from Ring B. Ring A also plays host to the Encke Division, discovered by Johann Encke in 1838. This division is much less conspicuous than the Cassini Division.

The rings themselves are comprised of countless fragments of varying composition that range in size from several metres down to a few microns. The differing brightness of the rings is a direct result of their composition, the prominent Ring B containing a higher abundance of ice and rock particles which reflect sunlight. Closer examin-

ation by space probes revealed that each of the main sections of the ring system is actually composed of many hundreds of narrow, closely-spaced ringlets.

Pioneer 11, which flew past Saturn in 1979, revealed the presence of the F Ring, which came under closer scrutiny during the subsequent Voyager flybys. It was found that the F Ring was only 100km (62 miles) wide and consisted of a number of intertwined strands. The reasons for this are not known.

Voyager also discovered a series of new ring systems. The D Ring is the innermost ring and extends from the C ring down to the cloud tops. The outermost G and E Rings are both extremely faint and lacking the ringlet structure predominant in the brighter rings.

A high resolution picture of the Cassini Division between the A and B rings of Saturn, taken from a distance of 13 million km (8 million miles).

49

Uranus

During the late evening of 13 March 1781 William Herschel was observing stars in the constellation of Gemini when he noticed an object that seemed anything but starlike in appearance. When he used higher magnifications, he saw that the object displayed a definite disc. Herschel knew that the stars were so far away that even with the most powerful telescopes they appeared as nothing more than points of light.

At first he suspected that the new object was a comet. Comets (or planets for that matter) travel around the Sun and consequently move through the sky against the background of stars. With this in mind, Herschel looked at the object again four nights later and indeed saw that it had changed position relative to the stars in the field of view.

Once mathematicians had plotted an orbit for the new object, it became clear that its path through the sky was nothing like one that would be expected of a comet. Its orbit was found to be almost circular, something that could only be explained if Herschel's object was in fact another planet orbiting the Sun.

Uranus, as the new planet came to be known, was found to be a distant world, orbiting the Sun at about twice the distance of Saturn. The planet had actually been seen several times during the preceding 90 years, in some cases by highly respected astronomers, yet it had always been mistaken for a star. However, these previous sightings did help astronomers to work out a more accurate orbit for Uranus. The fact that Herschel had recognized it as being something other than a star was a tribute to his skills as an observer. Uranus was the first planet to be discovered with a telescope, and its discovery effectively doubled the size of the known Solar System. Uranus was the first planet to be discovered in recorded history.

Uranus is a gas giant with a diameter of 50,800km (31,566 miles). It orbits the Sun once every 84.01 years at a mean distance of 2,869,600,000km (1,783,135,000 miles), spinning once on its axis every 17.24 hours. The planet can become visible to the unaided eye, although really clear, dark skies are required in order to see it without assist-

Two views of Uranus, one, on the left, in true colour is as a human eye would see it from Voyager. The picture on the right is a false colour image which reveals a dark polar hood surrounded by a series of lighter concentric bands which may be smog or haze.

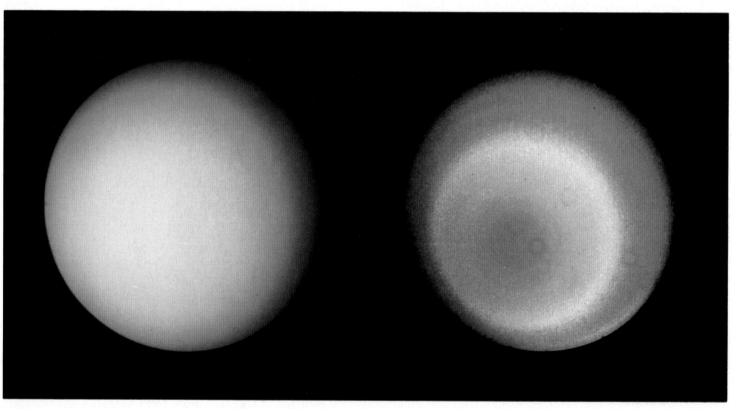

ance. When viewed through a telescope, Uranus displays a tiny, greenish disc.

Uranus has a very high axial tilt, the planet being tilted by 98° with respect to its orbit. This means that its axis of rotation is close to its orbital plane, resulting in its north and south poles pointing alternately towards the Sun. This leads to an unusual Uranian 'calendar' with each pole having a long 'summer' of 21 Earth years followed by an equally long period of darkness. It is not known for certain why the planet has such a high axial tilt, but it may have been the result of a cometary collision.

Inside Uranus

As with the other gaseous planets, Uranus is comprised mainly of hydrogen and helium. The temperature at its surface is somewhere in the region of −220°C. Prior to the visit of Voyager 2 in January 1986, little was known of the planet. Even the world's largest telescopes reveal little if any detail on Uranus' tiny disc, and it was left to Voyager to show us more of Uranus for the first time.

The thousands of images returned by Voyager increased our knowledge of Uranus dramatically. Clouds and mist were revealed, particularly prominent

Voyager 2 photo and accompanying geological map of Uranus' moon, Titania.

TITANIA

EXPLANATION

▨ BRIGHT MATERIAL	■ SMOOTH TERRAIN	▨ CRATERED TERRAIN
⊤ NORMAL FAULT BAR AND DOT ON DOWNTHROWN SIDE	⬭ BASIN OR CRATER RIM	
	+ SOUTH POLE	

being a huge cap of mist enveloping the sunward facing pole. Atmospheric bands running parallel to the equator were also photographed.

A rocky core, some 15,000km (9,321 miles) in diameter, is thought to lie at the centre of Uranus. The temperature in this region is around 7,000°C with the downward pressure of the material above it equalling 20 million times that of the atmosphere at the Earth's surface. Above the core is a 10,000km (6,214 miles) thick icy mantle containing methane, water and ammonia. Above this is the Uranian atmosphere which is made up primarily of hydrogen and helium. Also containing traces of other elements, it is the outer surface of this atmosphere that we see from Earth.

Voyager 2, high resolution photo of Uranus' satellite Miranda.

The Satellites of Uranus

Before the exploration of Uranus by Voyager 2 in January 1986 only five satellites were known to orbit Uranus, although the Voyager cameras increased this number to fifteen, as well as providing information on the five previously-known moons.

Titania is the largest of the Uranian satellites. Discovered by William Herschel in 1787 (the same year he located Oberon), it was seen to contain numerous impact craters, although fewer large craters than were seen on Oberon. Valleys, faults and fractures were seen across Titania's surface, many of which are thought to be quite young on the geological time scale. Several were seen to cut large craters in half and a heavily faulted region is thought to be caused by the crust fracturing as water froze below the surface.

Oberon contains many impact craters, some of which are surrounded by regions of bright ejecta. One particularly notable feature is a mountain which towers some 20km (12.4 miles) above the surrounding terrain.

The darkest of the five large satellites is Umbriel, a 1,172-km (728 miles) world that is almost uniformly covered in craters. Many large craters indicate that the surface is quite old, although there is a covering of dark material which appears to have hidden most of the topographic features apart from the craters. The source of this material is not known with certainty, although it did not necessarily originate from within Umbriel itself. Some scientists think that it may have come from a Uranian ring.

Miranda was the subject of some of the best images to be returned by the Voyager 2 probe. Surface features were photographed with a resolution of as little as 500m (1,640ft). A wide range of different features were seen including craters and plateaux, valleys and fractures, cliffs and canyons. Miranda, although only just under 500km (310 miles) in diameter, appears to contain examples of most of the different types of geological features found throughout the Solar System. There are two main areas of terrain, an older, crater-strewn landscape contrasting with a much younger type with fewer craters. It has been suggested that the complex and extremely varied surface of Miranda may be due to it having been broken up by a huge impact long ago. Rather then dispersing the fragments of the satellite, the impact was followed by a coalescing of the blocks of debris. Prior to the impact Miranda had a dense, rocky core overlaid by less-dense ice. Following the impact Miranda became a chaotic mix of different materials. The heavier, denser material will have gravitated towards the centre, leaving the lighter material nearer the surface.

Ariel contains few craters, most of these being around 50km (31 miles) or less in diameter. Large craters have disappeared from view. Ariel has been

the scene of much geological activity with many faults, fractures and similar features in evidence. The surface that we see today is thought to be quite young. Some features that appear to be volcanic in origin were imaged by the Voyager 2 cameras. Some form of viscous fluid seems to have once flowed across the surface, although just what material would be in liquid form at the low temperatures present on Ariel remains something of a mystery for scientists.

The other ten, smaller satellites are all, with the exception of Puck, less than 100km (62 miles) in diameter. They all orbit Uranus within the orbits of the five largest satellites. The first of the new discoveries was that of Puck which came

Voyager 2 photo and accompanying geological map of Uranus' moon Ariel.

ARIEL

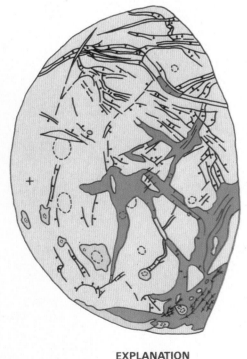

EXPLANATION

☐ BRIGHT MATERIAL	■ SMOOTH TERRAIN	☐ CRATERED TERRAIN

⊤ NORMAL FAULT
BAR AND DOT ON
DOWNTHROWN SIDE

➤ GROOVE
OR TROUGH

◌ BASIN OR
CRATER RIM

+ SOUTH POLE

The rings of Uranus taken by Voyager 2 as it passed into the shadow of the planet. Backlit dust is clearly visible between the rings.

to light in December 1985. Images of this satellite revealed a dark surface which reflects only a small percentage of the sunlight it receives. A number of craters were seen on the surface.

The presence of fifteen satellites orbiting Uranus (and there may well be more awaiting discovery), coupled with the fact that Jupiter (16), Saturn (21) and Neptune (8) all have large satellite families, illustrates that the outer Solar System is more heavily populated than was at first thought. Together with the large satellite families of the four gas giants there may well be further groups of asteroids orbiting the Sun in paths that keep them well away from the region of terrestrial planets and the main asteroid zone.

The Rings of Uranus
Just as Uranus had been discovered accidentally, its ring system was also found somewhat by chance. On 10 March 1977 Uranus passed in front of the faint star SAO 158687 in Libra, temporarily hiding the star from view. Events whereby a nearby celestial body

passes in front of a more distant object are known as occultations. Occultations of stars by planets are rare, and when they do occur they are carefully monitored. Observation of this type of occultation can provide a great deal of useful information about the planet.

Watching how the light from the star fades before it disappears behind the planet can tell us much about the density and opacity of its atmosphere. Also, by knowing the orbital speed of the planet, and by determining the length of time the star remains hidden, the diameter of the planet can be calculated. Our knowledge of the Uranian atmosphere and our estimates of the diameter of Uranus were a little sketchy to say the least, so it isn't surprising that the occultation of SAO 158687 was watched with particular interest.

Shortly before the star disappeared its light was seen to 'blink' five times, followed by a similar sequence of dimmings seen after the star reappeared. This could only mean one thing: Uranus had a ring system. The unexpected dips in the intensity of the light coming from the star were caused by the passage of a series of rings between the star and the observers.

Until this discovery only Saturn was known to have a ring system. The Jovian rings were revealed by the Voyager 1 cameras almost exactly two years later. Subsequent observation of the rings, together with the images returned by Voyager 2 in January 1986, has brought the number of known Uranian rings to eleven.

The ring system lies in the Uranian equatorial plane, circling Uranus between 42,000 and 52,000km (26,098–32,312 miles) from the centre of the planet. Their overall diameter is therefore in excess of 100,000km (62,138 miles). However, the rings are only very narrow, the widest being the outermost Epsilon ring. The width of the Epsilon ring varies considerably from 20 to 95km (12.5–59 miles), as does its distance from Uranus. So eccentric is the Epsilon ring that the distance of its inner edge from the Uranian cloud tops varies by as much as 800km (497 miles).

Most of the other rings have large eccentricities, with only a few of the rings being circular. The Epsilon ring is by far the largest of the Uranian rings, most of the rest being under 10km (6.2 miles) wide. The widths of many of the rings also vary. Unlike the rings of Saturn, which are very bright, those of Uranus are dark, the material from which they are formed reflecting only a small percentage of the light received from the Sun. The sizes of particles forming the Uranian rings is thought to range between a few centimetres and several metres in diameter.

As Voyager 2 passed into the shadow of Uranus and looked back at the ring system, it was seen that the gaps between the rings were not empty. Extensive bands of fine dust were seen scattered throughout the ring system. These particles are thought to be extremely small and may be the result of collisions between the larger particles forming the main rings.

All eight known rings of Uranus are visible in this photograph.

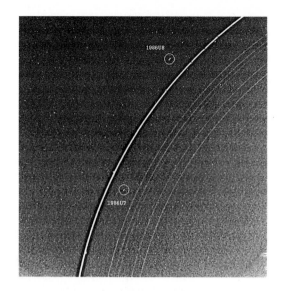

Shepherd Satellites

A pair of satellites, one on either side of the Epsilon ring, was photographed by Voyager 2. The edges of the ring lie very close to the resonances of the satellites. Collisions and other disturbances between the ring particles, and between the particles and the satellites, set up forces that govern the width of the ring. Surprisingly, no shepherd satellites were found for any of the other rings, possibly because any such satellites may be small, very dark and may have escaped detection by the Voyager 2 cameras.

Neptune

Following Herschel's discovery of Uranus, the Solar System was once more considered complete. However, it wasn't long before observations of Uranus revealed that the planet was wandering from its predicted orbit. Astronomers believed that these deviations were being created by the gravitational influence of another planet, orbiting the Sun at an even greater distance.

The young English astronomer and mathematician John Couch Adams decided to try and work out where the as-yet undiscovered planet might be. He used the deviations between the predicted positions of Uranus and the actual observed positions to try and locate its whereabouts. After many months of work he was satisfied that he had fixed the position accurately, and sent his results and predictions to George Biddell Airy, the Astronomer Royal at the time. Unfortunately, Airy failed to instigate a search for the hypothetical planet.

Unknown to Adams, the French mathematician Urbain Jean Joseph Le Verrier had calculated an almost identical position for the new planet using methods similar to those used by Adams. Le Verrier's predictions were sent to the Berlin Observatory where, on 23 September 1846, Johann Galle and Heinrich d'Arrest quickly located the planet in almost the same position as predicted by Le Verrier. Although it was Le Verrier's calculations that were eventually used to locate Neptune, both Adams and Le Verrier are credited for their mathematical achievements.

Voyager Encounters Neptune

Visible through large telescopes as a pale bluish disc, Neptune is the eighth planet out from the Sun. It is almost a twin of Uranus, being slightly smaller with a diameter of 49,500km (30,758 miles). It orbits the Sun every 164.79 years in an almost circular path. Its average distance from the Sun is 4,497 million km (2,794 million miles). Neptune is so far away that the radio signals from the Voyager 2 space probe, which flew past the planet in August 1989, took just over four hours to reach Earth.

As Voyager 2 passed within 5,000km (3,107 miles) of the Neptunian cloud tops, its cameras revealed a wide variety of features. Particularly prominent were bright polar collars and broad bands in different shades of blue, girdling Neptune's southern hemisphere. Also visible were bright streaks of cirrus cloud stretched out parallel to the equator. The Voyager cameras picked out the shadows of these clouds thrown onto the main cloud deck some 50km (31 miles) below.

A number of dark features were seen, the most remarkable of which was a huge oval storm cloud roughly the size of Earth situated at latitude 22° south. Named the Great Dark Spot, it is similar in many ways to the Great Red Spot on Jupiter. The GDS rotates in an anti-clockwise direction over a period of about 10 days. Observation revealed that this feature is actually a hole in the Neptunian clouds through which we can see the lower reaches of the atmosphere.

Cirrus-type clouds of frozen methane were seen forming and changing shape above and around the GDS. These clouds are created as the atmosphere, rich in methane, passes over the GDS, the pressures prevalent here causing the gases to cool and condense into clouds.

Another bright cloud feature was the 'Scooter', a bright area of frozen methane cirrus cloud located to the south of the GDS. This feature was given its name because it was seen to travel around the planet at a faster rate than the other clouds. The Scooter may be a plume rising up from the hydrogen sulphide cloud layer situated below the outer, visible cloud deck.

At latitude 54° south, another dark feature was seen. Nicknamed Dark Spot 2, this feature spins around a plume of bright, white methane clouds that appear to have condensed from rising methane-rich air. Dark Spot 2 was seen to rotate around the planet at a slightly faster speed than the Great Dark Spot further to the north, the difference in speeds being roughly 350km/h (217 mph).

The Satellites of Neptune

Prior to the Voyager 2 encounter, only two satellites were known to orbit Neptune, although the Voyager cameras brought this figure to eight. The largest of these is Triton, which has a diameter of 2,720 km (1,690 miles) and orbits Neptune once every 5.88 days in a retrograde direction (a direction opposite to Neptune's direction of axial rotation). Triton lies at a distance of 354,000km (219,971 miles) from the planet. Its surface contains many interesting features including fault lines and a bright polar region. This area reflects around 90 per cent of sunlight received.

Triton's density was found to be a little over 2 g/cm³, roughly twice the density of water. This suggests that Triton is made up of a mixture of rock and icy material.

Lava flooding appears to have taken place over large areas of Triton's surface. Once the lava froze in the intense cold it became subject to meteoritic bombardment and some of the areas imaged by the Voyager cameras show many impact

The route taken by Voyager 2 around Neptune and showing the accompanying ring arcs.

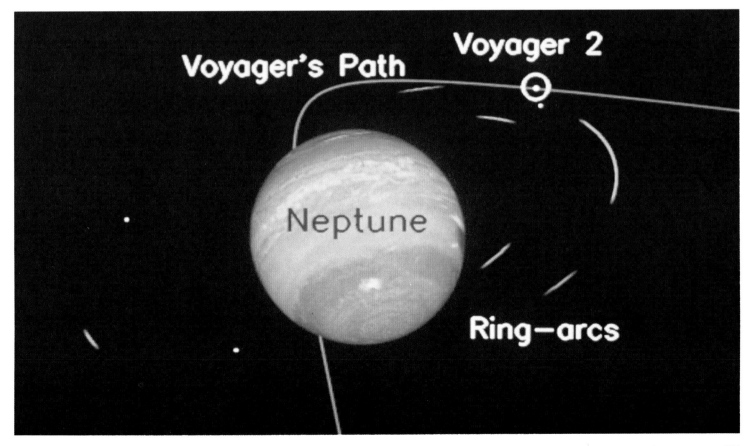

Three of Neptune's most prominent atmospheric features can be seen in this photograph. At the top is the Great Dark Spot accompanied by bright, white clouds that undergo rapid changes in appearance. To the south is the bright feature nicknamed the 'Scooter' because it rotates more rapidly around the planet than other features. Further south is Dark Spot 2.

craters, evidence which points to these regions being formed long ago. On the other hand, some lava flow regions displayed only a few craters, which suggests fairly recent formation, perhaps less than 500 million years ago. The lava which flowed across Triton's surface was not the molten rock-type material we see on Earth. Rather it was a mixture of water and other material, perhaps methane or ammonia. The lava must have had a melting point sufficiently low for flowing to take place, the surface temperature on Triton being such that pure water ice would be extremely hard and solid.

A large polar ice cap was detected on Triton, extending from the south pole to more than half way towards the equator. Dark streaks seen near the south pole may have been formed from liquid nitrogen being thrown up into the atmosphere to heights of several tens of kilometres, becoming frozen and then being 'blown' by gentle winds and deposited across the surface. The features seen certainly resemble wind streaks spotted elsewhere in the Solar System.

Another interesting surface feature were what seemed to be frozen lakes, which may have been formed by material

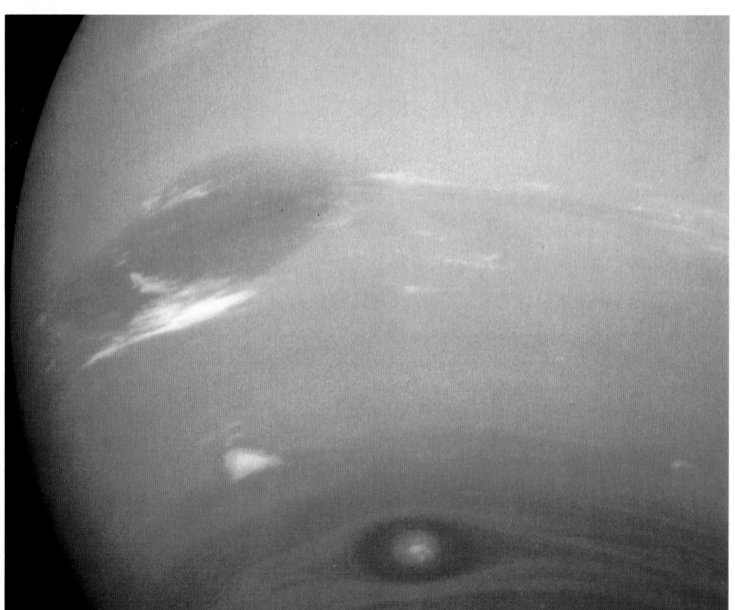

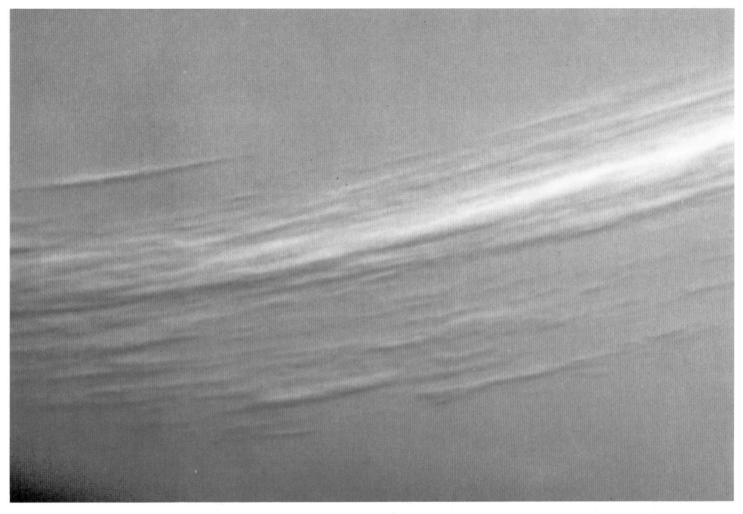

ejected from inside Triton and which froze in low-lying regions of the surface. Chemical reaction between solar radiation and the material on Triton's surface may have produced the pinkish colouring seen across Triton's southern hemisphere. The very thin atmosphere of Triton is comprised of methane and nitrogen with an atmospheric pressure at Triton's surface just 0.00001 that of the Earth at sea level.

Prior to the Voyager encounter, Nereid was thought to be the second-largest Neptunian satellite. However, Nereid has now been demoted to third place, following the discovery of a satellite (provisionally designated 1989 N1) some 400km (248 miles) across, which is about 60km (37 miles) larger than Nereid. Because Voyager came no closer than about 4,700,000 km (2,920,000 miles) to Nereid, the information we have gleaned

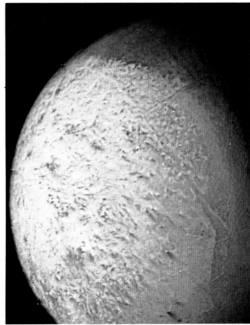

Above: This picture was taken two hours before Voyager's closest approach to Neptune. Fluffy white clouds can be seen floating high above the planet, casting cloud shadows, which have not been seen on any other planet. The heights of the clouds appear to be about 50km (30 miles).

Left: The bright southern hemisphere of Neptune's moon Triton.

59

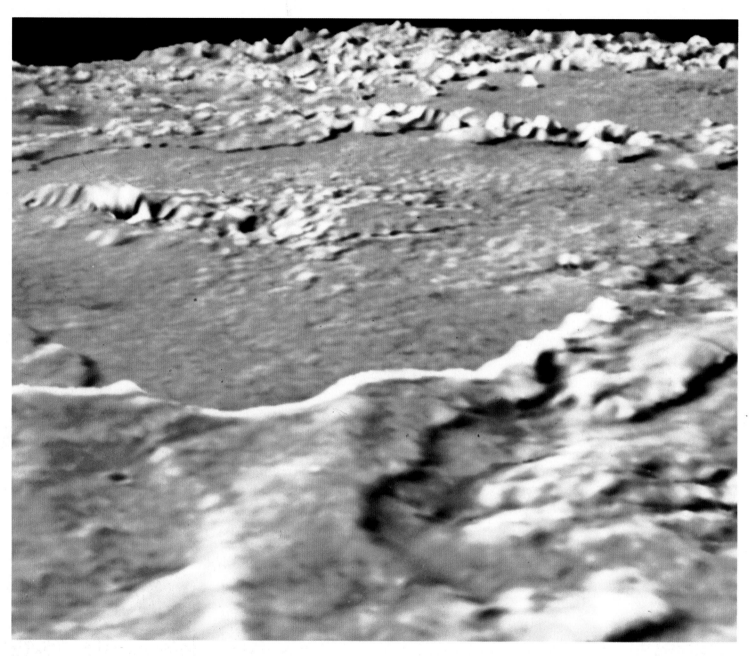

A computer-generated perspective view of one of Triton's caldera-like depressions.

from the images obtained of this satellite do little more than reveal its diameter and the reflectivity of its surface, which turns out to be around 12 per cent.

1989 N1 was found to orbit the planet at a mean distance of 117,600 km (73,075 miles). Had it been further out it might have been discovered through Earth-based observation long ago. Apart from Triton and Nereid, all the Neptunian satellites appear to be irregularly-shaped objects, all being too small for them to have formed uniform spheres.

The Rings of Neptune

One of the objectives of the Voyager 2 space craft, during its encounter with Neptune in August 1989, was to answer the question as to whether a ring system existed around the planet. Some Earth-based observations suggested that there might be at least a partial ring system, although the results were by no means conclusive. A number of observations during recent years of stellar occultations showed that on some occasions the light from the star dimmed momentarily on one side of the planet but not the other. If these dimmings were caused by material in the form of rings the light from the star would be expected to dim on both sides of the planet, and not just one as observed (see 'The Rings of Uranus'). To explain the observations astronomers came up with a theory of partial rings, or ring arcs, which did not completely circle the planet.

On 11 August 1989 the Voyager 2 cameras finally provided proof of the existence of a Neptunian ring system. Initially, the features discovered were thought to be ring arcs, although subsequent images showed that what the Voyager cameras had actually detected were brighter segments of rings that extended completely around the planet.

The ring system has been found to be comprised of two outer rings located at distances of 53,000 and 63,000 kilometres (32,940–39,155 miles) from Neptune. Both of these rings are very thin. A pair of satellites were found to orbit Neptune close to the rings, the presence of these 'shepherd' satellites keeping the ring particles confined. A third, much broader ring was found closer to the planet. This was so diffuse that it came to be known as the 'fuzzy' ring.

Voyager also detected a thin disc of particles extending from roughly midway between the two outer rings down towards the Neptunian cloud tops. This feature revealed itself on photographs taken 'looking back' at the ring system. The light from the Sun was backscattered by the fine particles in the disc, thereby increasing its visibility. This effect is similar to that observed by dirt on a car windscreen, which becomes very apparent if driving into the Sun when the dust scatters the Sun's light into your eyes.

The outer section of the disc has been christened the 'plateau' and is somewhat brighter than the rest. Discs of material like this are nothing new, similar features having been found in the ring systems of both Jupiter and Uranus.

Below: A detail of Neptune's rings.

Above: Neptune's ring system shown in two exposures each lasting nearly ten minutes.

Left: Neptune's satellite 1989N1 discovered by Voyager.

Pluto and Charon

The Discovery of Pluto

Pluto is the most recent planet to be discovered. Discrepancies in the movements of Uranus were still apparent even after the discovery of Neptune and searches for a trans-Neptunian planet were made by several individuals including the American astronomer Percival Lowell. Lowell's efforts proved negative, however, and he died in 1916 with the hypothetical planet escaping his detection.

Before he died, Lowell initiated the construction of a special wide-field camera to further the search. Several years later this was completed and a subsequent and extremely thorough photographic search for the trans-Neptunian planet was made by the young astronomer Clyde Tombaugh from the Lowell Observatory in Arizona. Tombaugh's efforts eventually paid off when the new planet came to light in 1930.

Pluto

Lying at a mean distance from the Sun of 5,900 million km (3,666 million miles) Pluto, the smallest planet, with a diameter of around 2,300km (1,429 miles), takes 248 years to travel once around our star. The orbit of Pluto is the most eccentric of any planet in the Solar System, and for twenty years of its orbit, Pluto comes inside the orbital path of Neptune. This last occurred in 1979 and until 1999, Neptune is actually the furthest planet from the Sun. Pluto's orbit is also more steeply inclined to the plane of the ecliptic than that of any of the other planets, its orbit being tilted at just over 17°.

Pluto is so tiny and distant that we are able to make out little if any surface detail on the planet and until quite recently even such details as mass, density and diameter were little more than educated guesswork. However, this was to change dramatically following the discovery of Pluto's satellite.

Charon

In 1978 the American astronomer James W. Christy was taking photographs of Pluto in an attempt to calculate future

occultations of stars by the planet. Observations of these events would help astronomers to estimate the diameter of the planet by timing the passage of the planet in front of the star. On one of the photographs Christy noticed a slight elongation of the Plutonian disc. A search through previous images revealed similar bumps on Pluto's image, leading astronomers to the conclusion that Pluto had a satellite.

The satellite was named Charon, and subsequent observation has shown that it orbits Pluto at a mean distance of 19,700km (12,241 miles) over a period of 6.387 days, a period identical to the rotational period of Pluto. This means that Charon and Pluto rotate synchronously, with Charon suspended permanently above the same point of the Plutonian surface. An observer on the Plutonian hemisphere facing Charon would see the satellite suspended in the

Below: The movements of Pluto and Charon are symmetrical in that Charon's period is the same as Pluto's rotation period. To an observer on Pluto, Charon would appear fixed in the sky. A transit of Pluto and Charon in 1978 revealed the existence of the latter.

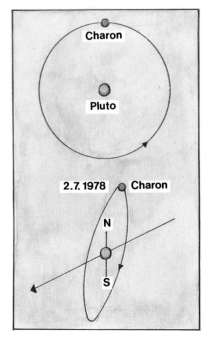

sky, neither rising nor setting and keeping the same angular distance from the horizon.

The discovery of Charon came at a very opportune time. Once the tilt of Charon's orbit around Pluto had been determined, it was realized that a rare alignment of the two bodies was about to occur. During the latter half of the 1980s Charon's orbital plane would be seen edge-on to the inner solar system. This meant that Charon and Pluto would undergo a series of mutual eclipses, each object periodically passing in front of its companion as seen from Earth. Such events are common with other planet-satellite systems, a good example being Jupiter. We often see Jovian satellites either crossing Jupiter's disc or passing behind the planet. However, as far as Pluto and Charon are concerned, such alignments are rare and only take place twice during Pluto's orbit around the Sun.

Timing the duration of each eclipse enabled astronomers to calculate the diameters of Pluto and Charon. Charon turns out to be around half the size of Pluto itself, so that the Pluto-Charon system should perhaps be more realistically regarded as a double planet rather than a planet and satellite.

Given the diameters and orbital details, the masses of the two objects could then be ascertained. Pluto's mass has been shown to be only 0.002 that of the Earth, and its density less than twice that of water. This value is consistent with a planet comprised of a mixture of rock and ice. This low mass also means that Pluto could not have caused the perturbations of Uranus and Neptune on which Lowell based his calculations. Only by coincidence was Pluto near the predicted position. The planet that Lowell predicted may still await discovery.

Part IV
Comets and
Meteors

Comets

The appearance of a comet in the sky in earlier times inspired both awe and fear and was often thought to presage momentous events. We now know that despite their visual splendour comets are rather tenuous members of our Solar System being composed largely of dust and gas. Nevertheless, comets are thought to have their origins at the time of the birth of the Solar System and their exploration will provide valuable insights into the processes whereby the planets were formed.

Oort's Cloud

The origins of comets are as yet unknown, but one of the more popular ideas was proposed in 1950 by the Dutch astronomer Jan Oort. He suggested the existence of a vast cloud of primeval material, situated beyond the orbit of Pluto and stretching out to a distance of around 2 light years from the Sun. Known as Oort's Cloud, it is said to surround the Solar System and is believed to be composed of particles left over from the formation of the planets. The clumps of icy material we see as comets originate from within Oort's Cloud.

Oort suggested that passing stars may produce gravitational perturbations which lead to the expulsion of material from Oort's Cloud. Some of these clumps will stray into interstellar space, although a number will fall towards the inner reaches of the Solar System. Slowly at first, but with increasing velocity, the frozen collection of dust and gas that will eventually reveal itself as a comet journeys from its home out beyond the orbit of Pluto to a rendezvous with the environs of the Sun.

Cometary Structure

The warmth exerted on a comet by the Sun increases as it approaches the inner regions of the Solar System. Its icy outer layers are vaporized and a cloud of material builds up around what will become the cometary nucleus. This cloud is known as the coma and it hides the nucleus of the comet from Earth-based observers.

The coma in turn will be acted upon both by the solar wind and radiation pressure as the comet gets closer to

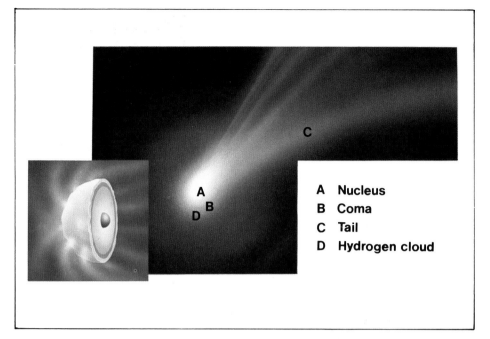

A **Nucleus**
B **Coma**
C **Tail**
D **Hydrogen cloud**

Above: The structure of a comet, the inset showing the 'icy conglomerated' model of a comet in which layers of ice surrounded a core, the whole being encased by an outer crust. The core may be silicate dust.

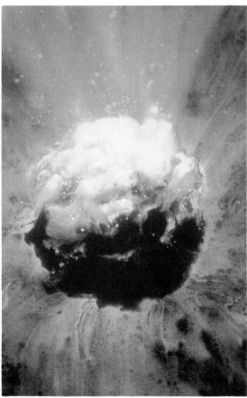

Left: An artist's impression of Halley's Comet as it is approached by Giotto.

Far left: Comet West, 9 March 1976, showing its dust tail (white) and its gas tail (blue).

perihelion, its closest point of approach to the Sun. Material is 'blown away' from the head of the comet, usually to form two tails; an ion tail, comprised of ionized atoms (atoms which have lost one or more electrons) and a dust tail.

Cometary orbits and periodicities. A
– Very long period; B – Long period
(Halley); C – Short period. (a)
Jupiter; (b) Neptune.

Because it is the constant force exerted
by the solar wind that produces the tails,
they are always seen to point away from
the Sun, the comet itself taking on the
appearance of a cosmic finger in the sky.

Gas released from water molecules,
which escape as the energy from the Sun
vaporizes the icy material within the
comet, forms a huge gaseous sphere
around the nucleus. Invisible to Earth-
based astronomers, this hydrogen envelope is millions of kilometres in diameter.

Once the comet begins to retreat from
the Sun the effects of the solar wind on

the comet diminishes, resulting in a
gradual disappearance of the tail
followed by the shrinking and disappearance of the coma. The icy nucleus is left
to carry on its lonely journey.

After its encounter with the Sun, a
comet may make its way back to Oort's
Cloud or, if it passes close to one of the
major planets, it may enter into a shorter,
closed orbit. In this case the gravitational
influence exerted by the planet may alter
the path of the comet, perhaps ensuring
that the celestial visitor may once again
grace our skies.

Cometary Orbits

Interactions with the Sun cause a comet to lose mass each time it passes through the inner regions of the Solar System, those with very short periods changing significantly and becoming quite faint. A notable example is Encke's Comet which, when first observed some 200 years or so ago, was reasonably conspicuous. However, its 3.3 year orbit (the shortest known orbital period for a comet) has led to an almost continuous barrage by solar energy, leading to the loss of most of its material into space.

Comets are usually named after their discoverers, although there are exceptions, one of which is Encke's Comet. Encke's Comet was seen by several observers between 1786 and 1818. The German mathematician Johann Encke deduced that these sightings were all of the same object, predicting its return in 1822. The comet indeed reappeared and was named in Encke's honour.

Other comets take much longer to orbit the Sun, an example being Halley's Comet (see 'Halley's Comet in History'). Some comets have periods that are so long we cannot measure them accurately. As a result, their appearances cannot be predicted and they are apt to take astronomers by surprise.

A reproduction of one of the original plates of Comet Halley taken at Helwan, Egypt, on 25 May 1910. The yellowish colour is a result of the processing.

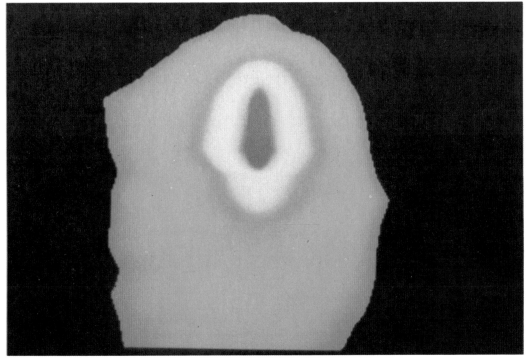

Left: Comet IRAS-Araki-Alcock of 1983, a small comet which came particularly close to the Earth. This false colour image is coded from red (brightest) to blue (faintest).

Catastrophic Encounters

The material within comets can become rapidly depleted if it passes very close to the Sun. Those which do are termed 'Sungrazers' and, as the comet passes perihelion, the original tail is often destroyed, a new one forming to take its place. Because cometary tails are produced through the evaporation of material from the nucleus, comets are short-lived by cosmic standards. Some comets have even disappeared, a notable example being Biela's Comet. Biela's Comet was discovered on three separate occasions: by Montaigne in March 1772, by Pons in November 1805 and by Wilhelm von Biela in February 1826. Von Biela calculated that all three sightings were of the same object, and that it would return again in 1832, which it did. It was missed during the return of 1839, although it was seen again in 1845. In early 1846 the comet was seen to split up into two parts. Both parts were seen again in 1852 although the pair went unnoticed in 1859 and again in 1865–66. It was due to return again in 1872. Although there was no sign of it, in November 1872 an extremely active meteor shower was seen. Additional meteor displays were observed at the predicted 1885, 1892 and 1899 returns, although subsequent displays, which occur on an annual basis, are nowhere near as active. The particles producing the Andromedid meteor shower have been proved to come from the now-defunct Biela's Comet. Biela's Comet exemplifies the connection between comets and meteors, meteors being regarded as cometary debris.

Halley's Comet in History

Halley's Comet is the best known of the periodical comets and records of sightings go back over 20 centuries. It orbits the Sun once every 76 years, this orbital period enabling many to have seen Halley's Comet twice during their lifetimes.

In 1705, the English astronomer Edmond Halley began to calculate the orbital motions of 24 bright comets seen between 1337 and 1698. He noticed that the orbits of the comets observed in 1531 and 1607 and a bright comet he himself had seen in 1682 were similar. The intervals between the sightings were also roughly identical at around 76 years. He suggested that these sightings were all of the same object and that it would reappear in 1758.

Halley died in 1743. Although professional astronomers searched for the comet as the date forecast by Halley drew near, it was the Dresden amateur astronomer Johann Georg Palitzsch who spotted it first on Christmas Day, 1758. The comet was named after Halley in

Comet Kobayashi-Berger-Milon, 1975. Four gas streamers are visible in the tail, as is the spiral galaxy Messier 106 in Ursa Major.

honour of his correctly predicting its return.

Dates of visits of the comet prior to 1531 have been calculated by plotting its orbital motion back in time, taking the gravitational effects of the planets into account. Some of the very earliest dates have been verified by checking against ancient astronomical records.

The first definite appearance of Halley's Comet was in 240 B.C. although the 12 B.C. appearance is the first about which we have detailed information. The most famous return was that of 1066 when its appearance was taken as a bad omen by the Saxons and, in particular, by Harold, the last of the Saxon kings. Following the invasion of England by William of Normandy, Harold died at the Battle of Hastings in October, 1066. The Bayeux Tapestry shows the comet suspended above Harold who is seen tottering on his throne as his courtiers look on in awe and terror.

After the comet's predicted return in 1758–59, and the discovery of Uranus in 1781, astronomers were able to plot its orbit with more accuracy by taking the gravitational effects of Uranus into account. Long before its scheduled return in 1835, attempts were made to calculate when and where it would reappear. As in 1758–59, the search for the returning comet started early, almost a year in fact before it was due to sweep through the inner Solar System. The first sighting was made in August, 1835 by astronomers at the Collegio Romano Observatory.

All our information about the 1835 appearance is in the form of sketches and visual descriptions. Photography was yet to make an impact on astronomy. In 1910, however, astrophotography provided the most comprehensive and detailed study of Halley's Comet up to that time.

The 1910 visit was the third predicted return, and was awaited eagerly by astronomers. The first astronomer to detect the returning comet was Professor Max Wolf at Heidelberg in Germany on a photograph taken on the night of 11/12 September, 1909. The comet was close to its expected position, although it didn't become visible to the naked eye until well into 1910. Before this, another bright

Halley's Comet, photographed from Cairns, Australia, on 24 April 1986.

comet made an unexpected appearance. The Great Daylight Comet, first seen by diamond miners in South Africa on 13 January, 1910, became a brilliant object, its tail stretching almost a fifth of the way across the sky by the end of January. This object became visible to the naked eye even in broad daylight, hence its name.

Halley's Comet passed between the Sun and Earth on 18 May, although no trace of the nucleus could be seen as the comet crossed the solar disc suggesting that the nucleus must be tiny and the gas around it very thin. During this time, it was thought that the Earth might pass through the tail, although there is no evidence that this actually occurred.

The comet then made its way back to the outer regions of the Solar System, and was last seen beyond the orbit of Jupiter. Little did astronomers expect that on its next visit Halley's Comet would be greeted by an armada of space probes and would be studied in more detail than ever before!

Halley in Close-Up

The return of Halley's Comet in 1985–86 provided astronomers with their best ever chance of exploring a comet. Because the orbital path of Halley's is known to a high degree of accuracy it was possible to plan missions by unmanned space probes to rendezvous with the comet.

In all, five space probes were sent to examine the comet. Two of these were the Soviet Vega probes, launched in December 1984. Prior to their encounters with Halley's Comet the Vega probes flew past Venus, dropping balloons into the Venusian atmosphere. They passed the comet on 6 March and 9 March, 1986 at distances of 8,890km (5,524 miles) and 8,030km (4,989 miles) respectively. Among the equipment they carried were cameras. The name Vega is derived from the Russian VEnera and GAlley, the latter being the Russian pronunciation of Halley.

False-colour image of Halley taken by the Soviet probe Vega 2.

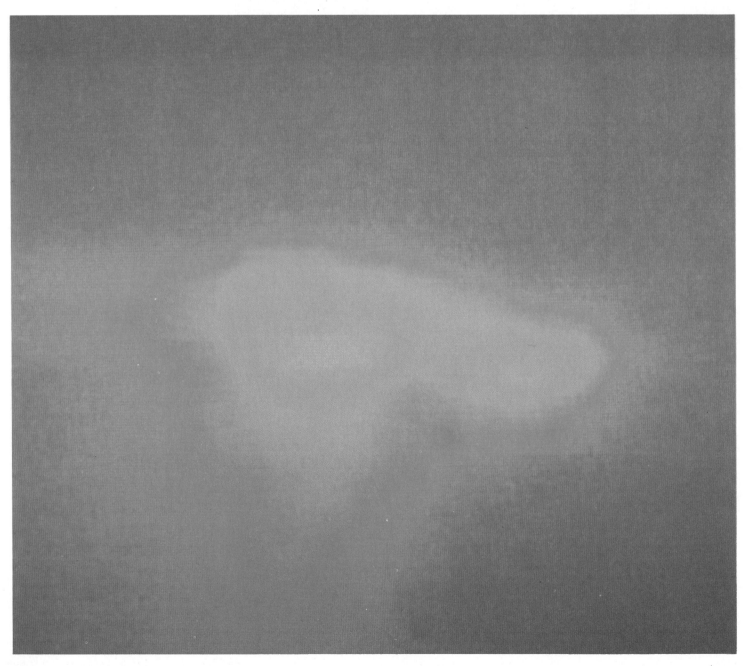

The Giotto spacecraft during integration at the Interspace facility in Toulouse. The camera is white, looking down, the star mapper in red (looking up) and the Neutral Mass Spectrometer is the red plate. The silver spheres are two of the four hydrazine tanks.

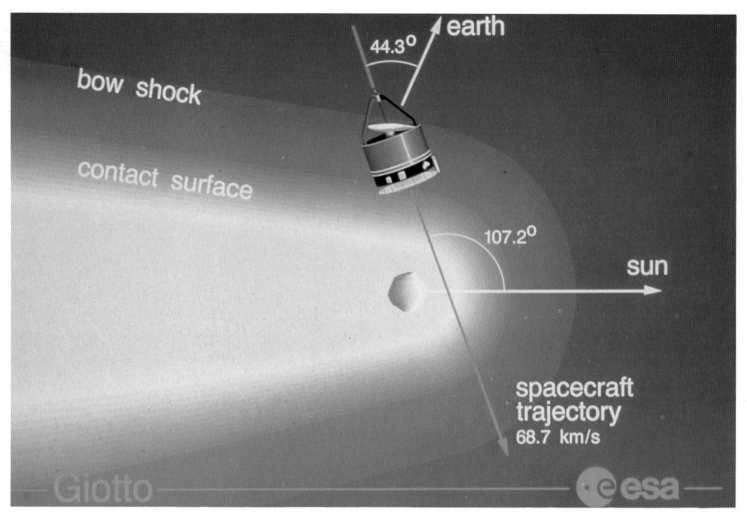

A schematic of the encounter geometry of Giotto with Halley. Sizes are not to scale.

The two Japanese probes, Sakigake and Suisei, carried out their investigations from greater distances. Sakigake, launched in January, 1985 flew by the comet on 11 March, 1986 at a distance of 6,900,000km (4,287,000 miles), its main purpose being to investigate the interaction between the solar wind and the comet at a large distance from the comet. One of the main aims of Suisei, launched in August, 1985, was to investigate the growth and decay of the corona of Halley's Comet. A comet's corona is a huge cloud of rarefied hydrogen gas thought to completely surround the head of the comet and which can only be studied from space probes. Suisei flew past the comet on 8th March, 1986 at a distance of 151,000km (93,830 miles).

By far the most successful of the probes to Halley's Comet was the European Giotto, named in honour of the Italian painter Giotto di Bondone and launched towards the comet on 2 July 1985. Giotto was cylindrical with a length of 2.85m (9.35ft) and a diameter of 1.86m (6.1ft). It carried a camera for photographing the nucleus and inner coma of Halley's Comet. Giotto flew within 610km (379 miles) of the nucleus on 14 March 1986 at a speed of over 65km/sec (40 miles/sec).

The data collected by Giotto was immediately transmitted back to Earth via a special antenna mounted on the end of the space probe facing away from the comet. This information was received back on Earth by the 64-metre antenna at Parkes ground station, Australia. A special shield protected the Giotto probe from impacts by dust particles during its passage through the comet's head.

The exploration of Halley's Comet by space probes was a truly international effort, the images and measurements obtained by the Soviet Vega craft helping scientists to precisely target Giotto. From Earth, the nucleus of a comet is hidden from view by the material surrounding it, and it wasn't until the Vega images were received that its position was established, and the subsequent trajectory of Giotto determined. During the encounter, all the instruments performed well, although disaster struck immediately before closest approach to the nucleus. A dust particle hit Giotto, temporarily knocking the spacecraft, and hence the antenna, out of alignment with Earth. After the encounter it was found that around half of the scientific experiments had suffered damage, although scientists were able to redirect the craft and put it on a course back for Earth. If all goes to plan, Giotto will be placed into a new orbit that will allow it to intercept and explore another comet, probably Comet Grigg-Skjellerup, which it would encounter in July 1992.

The Giotto camera showed the nucleus of Halley's Comet to be a small and irregularly-shaped chunk of ice, measuring some 15km (9.3 miles) long by 8km (5 miles) wide, coated by a layer of very dark material. This layer is thought to be composed of carbon-rich compounds and reflects just 4 per cent of the light it receives from the Sun. The nucleus of Halley's Comet is one of the darkest objects known.

The gas and dust that gives rise to all the cometary phenomena we see, including the coma and tail, was found to be escaping from the nucleus through vents

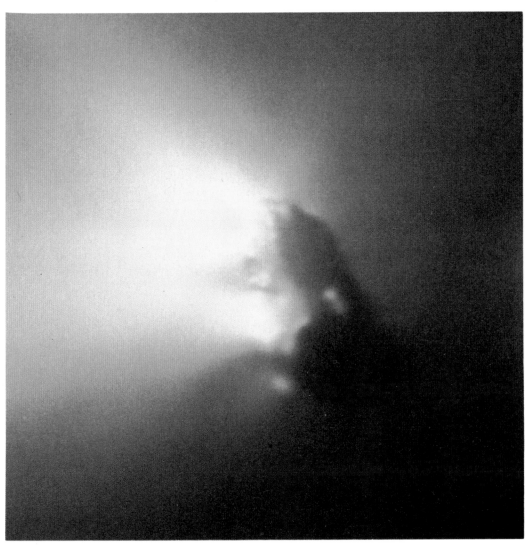

This image of the nucleus of Halley's Comet is composed of seven images taken by the Halley Multicolour Camera on Giotto.

An artist's impression of the Comet Rendezvous Asteroid Flyby (CRAF) spacecraft as it ejects a penetrator towards the nucleus of a comet. CRAF is the first mission of the Mariner Mark II series.

in the outer dust/ice layer. These vents become active when exposed to the Sun, ceasing to expel material when plunged into darkness as the nucleus rotates. The force of the jets of material escaping from the nucleus affect the comet's motion around the Sun. Halley's Comet was actually several days late in reaching perihelion this time because of the jet-like effects of the matter expelled.

As with the surface of the nucleus, the dust thrown off by the comet was found to be very dark and may have indeed come from the surface rather than the interior. Over three-quarters of the gas ejected from the nucleus was found to be water vapour, which appears to make up over 80 per cent of the nucleus. This was the first time that water had been positively identified in a comet in spite of the

fact that cometary nuclei were thought to consist of a mixture of dust and water ice.

Although our exploration of Halley's Comet has taught us a great deal about comets in general, there is still much to learn about these ghostly visitors to our region of the solar system. Future exploration by space probe of comets will include rendezvous missions, during which a probe will position itself close to a cometary nucleus for a prolonged period of time and perhaps send a lander to the surface of the nucleus. The possi-

bilities of such a mission are already being examined by NASA. Known as Comet Rendezvous and Asteroid Flyby (CRAF), scientists would be able to undertake close-up exploration of both asteroids and comets. Sample return missions, by which we will be able to examine first-hand material plucked from the heart of a comet, are also a possibility for the future. These are ambitious plans, and who can rule out the idea of a manned mission to Halley's Comet.

An image-processed optical photograph of the head of Halley's Comet, derived from four exposures made on 25 May 1910, at Helwan, Egypt. Computer enhancement reveals a small jet extending to the left from the centre of the nucleus and visible as small kinks in the inner red contours.

Right: A painting by artist Paul Doherty of a bright fireball.

Far right: Artist Paul Doherty's impression of the Leonid shower.

Meteors and Meteorites

If you look up into a clear, dark sky you may see a rapidly moving streak of light against the starry background. This is a meteor, and is created by a tiny particle of dust, or meteoroid, entering the atmosphere from interplanetary space as a result of the Earth's gravitational pull. The speed at which the particle enters the atmosphere can be quite high, as much as 70km/sec (43.5 miles/sec), resulting in it burning up through friction with air molecules and producing a streak of incandescence.

Countless numbers of meteoroids are travelling around the Sun, although they are much too small and faint to be seen until they enter the Earth's atmosphere. The friction set up destroys the particle long before it reaches the ground. Meteors have been known since antiquity, although it wasn't until the beginning of the nineteenth century that their true nature became fully understood.

Meteor Showers

Meteors can be seen on any clear night of the year, and can enter the atmosphere from any direction. These are known as sporadic meteors. However, at certain times of the year we witness concentrations of meteors, all of which appear to come from the same point in the sky. These so-called shower meteors are associated with comets, typical examples being the Taurids, which are linked with Encke's Comet and the Leonids, which are associated with Comet Tempel-Tuttle.

Comets shed material which eventually becomes scattered all along their orbital paths. Repeated passages through the inner Solar System causes comets to regularly lose material into space. The Earth passes through the orbital paths of certain comets at particular times of the year, at which times larger than average numbers of particles are drawn into the Earth's atmosphere producing meteor showers.

Particles emanating from a comet travel in parallel orbits to that comet. As a result, when such particles enter the atmosphere, the meteors seem all to radiate from the same point in the sky. This point is known as the 'radiant'. A

Wolfe Meteorite Crater in Western Australia. It is less well-known than the Arizona Meteor Crater but no less interesting.

similar effect is seen when looking up a long, straight railway line from a point above the tracks. The rails will seem to meet at a point near the horizon, which may be termed their apparent 'radiant'.

Meteor showers are named after the area of sky containing their radiants. For example, the Taurids radiate from the constellation Taurus while the Leonids radiate from a point in the constellation of Leo (close to the star Gamma Leonis).

Meteorites

Meteorites are larger objects which enter the atmosphere and survive the journey to the Earth's surface without being destroyed. There are three main types of meteorite. Iron meteorites make up around 5 per cent of all observed falls. Iron meteorites are further classified into around a dozen groups. Much more common are the stony meteorites which account for around 95 per cent of

observed falls. The ranks of stony meteorites are made up of chondrites and achondrites, each class having its own characteristics. The rarest group of all is the stony-irons which are split up into two further groups.

Descriptions of meteorites have come down to us from ancient times, one of the oldest reliable accounts dating from 654 B.C. and relating to a group of meteorites falling near Rome. Yet the origin of meteorites has been disputed until relatively recently. The idea of these objects coming from space had little support, and in 1795, when a 25kg (55lb) meteorite fell at Wold Cottage in Yorkshire, it was suggested that the object was actually a stone hurled out of an Icelandic volcano! However, in April 1803 the famous French physicist Jean Baptiste Biot showed that a number of meteorites which fell over the town of L'Aigle in France came from the sky. Biot's research moved meteorites from the realms of geology to those of astronomy.

Meteorite Craters

Fortunately, very large meteorite falls are rare, for these events can cause a great deal of damage around the impact site. The huge numbers of impact craters seen on the Moon and other planets bears testimony to this. The best known example of a terrestrial meteorite crater is the Arizona Meteorite Crater, a large depression over a kilometre (0.6 mile) in diameter and almost 200m (656ft) deep, formed during a meteorite impact that took place thousands of years ago. Many other large meteorite craters have been found, including those at Lac Coutoure in Quebec and Manson in Iowa.

Many museums have collections of meteorites, the largest meteorite on display being the 31-tonne object at Hayden Planetarium in New York. The largest known meteorite, weighing around 60 tonnes, lies where it fell in prehistoric times at Hoba West, near the town of Grootfontein in South West Africa.

Origins of Meteorites

Meteoroids are comparatively large objects that travel around the Sun in Earth-crossing orbits. There are many meteorite falls per year, and the study of meteorite paths prior to atmospheric entry has revealed that many could have originated within the asteroid belt, coming about as the result of collisions between larger bodies. These findings are borne out through examination of meteorites themselves which places their ages at around 4,500 million years, and it is thought that they originated at the same time as the other members of the Solar System.

The Hoba West Meteorite, near Grootfontein, South Africa. The date of the fall of this iron meteorite is not known, but it is almost certainly prehistoric.

Part V
The Future of
Planetary Exploration

The Voyager 2 encounter with Neptune in August 1989 brought to a close the second phase in our exploration of the Solar System. The first phase involved initial reconnaissances of the planets by spacecraft, including the lunar surveys prior to the manned landings and the Mariner series which gave us our first glimpses of Mars, Mercury and Venus. Phase 1 culminated with the achievements of Pioneer 10 and 11 at Jupiter and Saturn. The vastly more sophisticated Voyager craft took the work of the Pioneer vehicles a great deal further.

But what of the future? At the moment, no further flights are planned to Mercury, Uranus or Neptune, and Pluto has yet to be considered. However, the Soviets plan a series of exploratory missions to Mars over the next decade, prior to a manned landing early in the 21st century. These missions will include both a balloon to carry instruments across the Martian surface and a sample return mission which may provide scientists back on Earth with our first specimens of Martian soil by 1996.

The Americans also plan to take a closer look at Mars with their Mars Observer mission, due for launch in 1992 and intended to build on the success of the Mariner 9 and Viking 1 and 2 missions of the 1970s. The scientific objectives of the Mars Observer probe include a survey of the Martian climate, including seasonal variations, detection of water and the observation of cloud composition, to be carried out during a 2-year (or 1 Martian year of 687 days) mission.

Cometary exploration is also planned in the near future. The European Space Agency is planning a Comet-Nucleus Sample-Return (CNSR) mission, due for launch at the beginning of the 21st century. CNSR involves the launch and operation of a spacecraft capable of making a rendezvous with a comet, landing on the surface of its nucleus, retrieving samples and returning them to Earth. Several comets have already been earmarked as possible targets for the CNSR mission, all of which would require a launch in either 2001 or 2002.

CNSR will provide us with our first sample of cometary material and will allow us to study what is thought to be the raw material of the Solar System, material which has remained relatively unchanged since its formation billions of years ago. Collaboration with NASA has been proposed, the NASA Solar System Exploration Committee having also recommended a comet-nucleus sample-return mission for the early-21st century.

Scientists have also devised exciting new exploratory missions to both Jupiter and Saturn together with further studies of Venus (all detailed below) and even now space probes are making their way towards what will be the first of a new wave of planetary encounters, a wave that will herald the start of the third phase in our exploration of the Solar System.

Galileo

Launched in October 1989 from the cargo bay of Atlantis during mission STS-34, the 2-tonne Galileo craft, named after the Italian astronomer Galileo Galilei, who discovered the four largest satellites of Jupiter, is currently heading for a rendezvous with Jupiter in 1995. This rather long flight time is due to the craft trajectory which took it past Venus in February 1990, the Venusian gravity being used to accelerate the craft back towards Earth. After making observations of the lunar farside, the Earth's gravity in turn will propel Galileo on a 2-year journey around the Sun towards another Earth rendezvous in December 1992.

During the interval between the two Earth encounters, Galileo will pass through the asteroid belt, approaching to within 1000km (621 miles) of asteroid Gaspra in October 1991. During the second Earth-encounter Galileo will make a photographic survey of the lunar north pole to try and determine whether ice exists in this region. A second asteroid encounter, in this case with Ida in August 1993, will follow, prior to a December 1995 arrival at Jupiter. The asteroid encounters are intended to explore both composition and surface geology.

The broad objectives of the Galileo mission are twofold and involve both an orbiter and a descent probe. Five months before the arrival at Jupiter, an atmos-

Far left: The Mars Rover Sample Return (MRSR) mission will send a lander to Mars to collect soil samples for transport to Earth. It will be a precursor to a manned mission.

The probe Galileo in the cargo bay of the Space Shuttle Atlantis.

The burn of the Inertial Upper Stage (IUS) rocket carries the Galileo spacecraft away from the Space Shuttle.

pheric probe will be deployed. This will be targeted towards a point a few degrees above the Jovian equator. The probe is encased in a protective heat shield which will be needed to prevent it burning up as it hits the outer layers of the Jovian atmosphere at a speed of almost 175,000km/h (108,743 mph). The probe will be slowed down to subsonic speed in less than 2 minutes after which the covers will be released, a parachute will be deployed and the on-board instruments and radio transmitter will go to work providing scientists with the first direct sampling of the Jovian atmosphere.

Information gathered by the probe during its intended 75-minute operational period include details of the pressure, density and temperature of the atmosphere, its chemical composition including the ratio of hydrogen to helium and the detection and examination of lightning. By the time its investigations come to an end the probe will be subjected to an atmospheric pressure equivalent to around 20 times that at the Earth's surface.

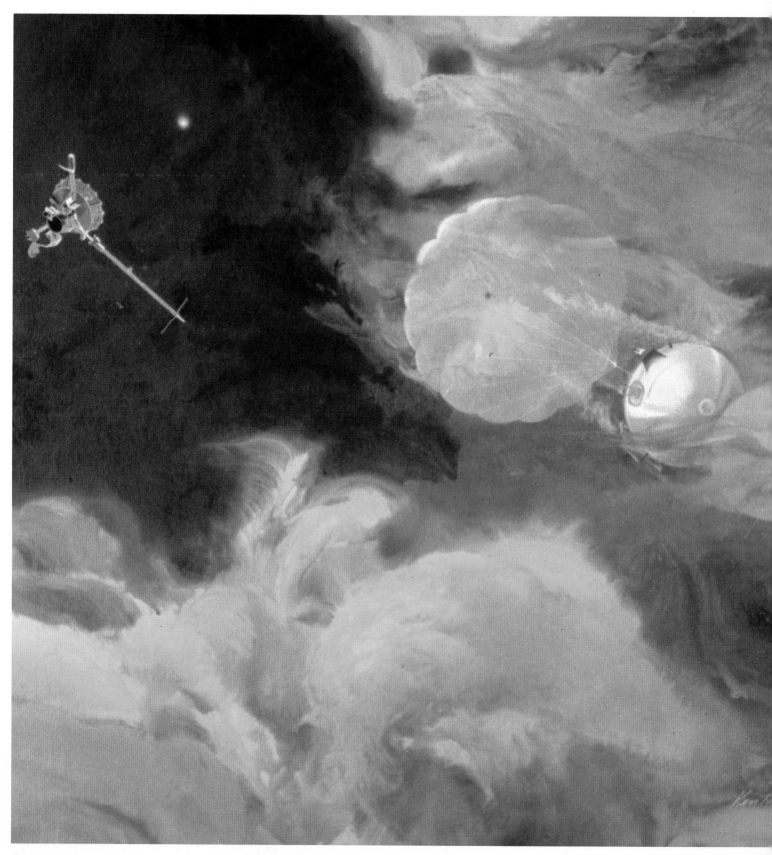

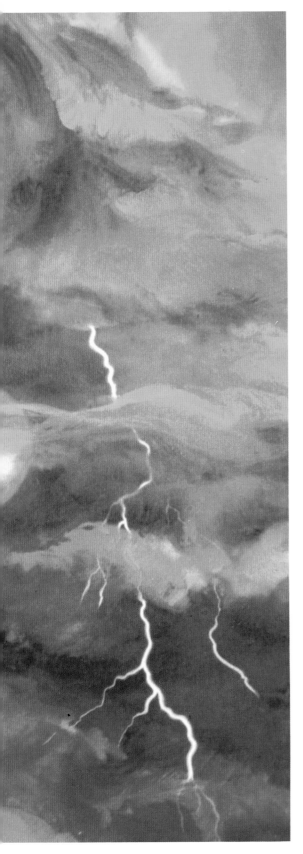

As it approaches Jupiter, and a few hours before the probe enters the Jovian atmosphere, the orbiter will pass within 1,000km (621 miles) of Io, the gravitational field of which will be used to slow the probe down prior to entering into orbit around Jupiter. As well as relaying the data from the atmospheric probe back to Earth, the orbiter will go on to perform its own programme of investigation, including high resolution imagery of the Galilean satellites, a study of the Jovian ring system and a detailed investigation of the Jovian atmosphere and magnetosphere. The Galileo orbiter will be the first craft to pass through the Jovian magnetotail which cannot be observed from Earth since it lies behind the planet. The orbiter will make 11 orbits of Jupiter, approaching to within 200km (124 miles) or so of some Jovian satellites and returning data for around 22 months.

Magellan

The Magellan craft was launched to Venus from the shuttle Atlantis in May 1989 during mission STS-30 and will reach its destination in August 1990 after a 15-month flight. The Magellan craft was the first planetary probe to be launched from the space shuttle, and the first to be launched for over a decade.

Following in the footsteps of earlier missions to the planet, Magellan will be

This artist's drawing shows the Galileo probe as it descends into Jupiter's stormy atmosphere to take samples of cloud layers.

The Magellan spacecraft, attached to an Inertial Upper Stage (IUS) rocket, is carried into low Earth orbit by the Space Shuttle. The probe will travel one and a half times around the Sun before it arrives at Venus 15 months after launch.

mapping the Venusian surface, but with a higher resolution than obtained on previous missions. Up to 90 per cent of the surface will be mapped during Magellan's 243-day (1 Venusian day) mission. The craft will orbit Venus once every 189 minutes, its distance from the planet varying between 250 and 8,000km (155–4970 miles). The information gained will allow us to formulate a global geological analysis of Venus comparable to those obtained for the other terrestrial planets.

Magellan undergoing testing at Martin Marietta Astronautics in Denver. As a cost saver, several major pieces of the craft are spares from other missions.

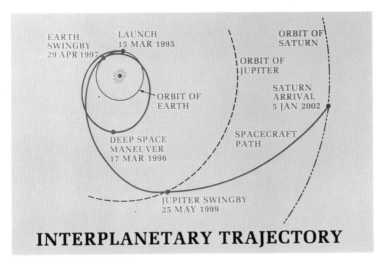

INTERPLANETARY TRAJECTORY

Labels: EARTH SWINGBY 29 APR 1997 · LAUNCH 15 MAR 1995 · ORBIT OF SATURN · ORBIT OF JUPITER · ORBIT OF EARTH · SATURN ARRIVAL 5 JAN 2002 · DEEP SPACE MANEUVER 17 MAR 1996 · SPACECRAFT PATH · JUPITER SWINGBY 25 MAY 1999

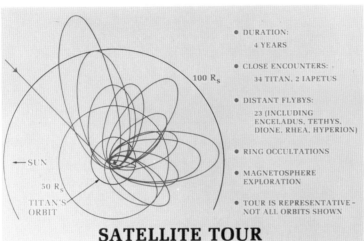

SATELLITE TOUR

- DURATION:
  4 YEARS
- CLOSE ENCOUNTERS:
  34 TITAN, 2 IAPETUS
- DISTANT FLYBYS:
  23 (INCLUDING ENCELADUS, TETHYS, DIONE, RHEA, HYPERION)
- RING OCCULTATIONS
- MAGNETOSPHERE EXPLORATION
- TOUR IS REPRESENTATIVE - NOT ALL ORBITS SHOWN

Labels: 100 R_S · SUN · 50 R_S · TITAN'S ORBIT

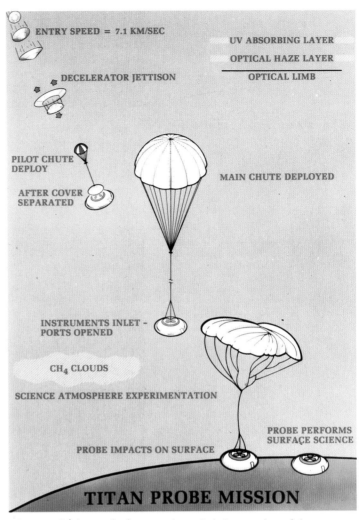

TITAN PROBE MISSION

Labels: ENTRY SPEED = 7.1 KM/SEC · DECELERATOR JETTISON · UV ABSORBING LAYER · OPTICAL HAZE LAYER · OPTICAL LIMB · PILOT CHUTE DEPLOY · AFTER COVER SEPARATED · MAIN CHUTE DEPLOYED · INSTRUMENTS INLET - PORTS OPENED · CH_4 CLOUDS · SCIENCE ATMOSPHERE EXPERIMENTATION · PROBE IMPACTS ON SURFACE · PROBE PERFORMS SURFACE SCIENCE

Cassini

This joint ESA/NASA mission, provisionally set for a launch in 1996, will despatch a probe towards the planet Saturn. As with the Galileo mission to Jupiter, the Cassini flight involves both an orbiter and an atmospheric probe. The orbiter will carry out a detailed, four-year survey of the Saturnian system, including close approaches to many of the icy satellites together with up to 40 flybys of Titan. Examination of the structure and composition of the rings will also be carried out as will a study of the planet itself. Cloud properties, atmospheric composition, winds and temperatures will all come under the scrutiny of Cassini.

The Cassini probe will reach Saturn in late 2002 (following a 1996 launch) its trajectory taking in flybys of the asteroid Maja in 1997 and Jupiter in February 2000. The first phase of the mission following arrival at Saturn is the targeting of a probe towards Titan. This probe will enter Titan's atmosphere on the daylight side at a speed of around 7km/sec (4.3 miles/sec). An aerodynamic decelerator will slow it down to about 270m/sec (886ft/sec) at an altitude of 175km (109 miles), following which a parachute will allow for a slow descent to Titan's surface. The descent will take up to 3 hours during which time analysis of the atmosphere will be carried out. Although the instruments are not designed to survive the 4–6m/sec (2.5–3.7ft/sec) impact, the probe may survive long enough for the collection and transmission of data from the surface. All data will be returned to Earth via the Cassini orbiter.

* * * *

An artist's impression of the NASA/ESA project, Cassini. NASA will provide the orbiter and ESA the smaller probe which will descend through the atmosphere of Saturn's moon Titan. The Titan probe is called Huygens after the Dutch astronomer who discovered Titan and the Saturnian rings in 1656.

After its near-fatal encounter with Halley's Comet in March 1986, the European Giotto probe entered into its own orbit around the Sun. However, the European Space Agency carried out a number of tests in 1989 which will pave the way for the reactivation of Giotto. Following its close approach to Earth in July 1990 the craft will, if all goes to plan, be put into a new orbit that will allow it to intercept another comet. This will probably be Comet Grigg-Skjellerup, which it would encounter on 10 July 1992.

Comet Grigg-Skjellerup was discovered in 1902 by John Grigg from Thames, New Zealand and independently by J. F. Skjellerup from the Cape of Good Hope, South Africa. This comet has one of the shortest orbits known, travelling once around the Sun every 5.1 years. Giotto's encounter with Halley's Comet in 1986 taught us a great deal about comets. If the encounter with Grigg-Skjellerup goes ahead successfully, we will doubtless learn even more about these ghostly and ethereal members of the Solar System.

Right: A diagram showing the path of Halley's Comet around the Sun, together with the orbits of Mars, Earth, Venus and Mercury.

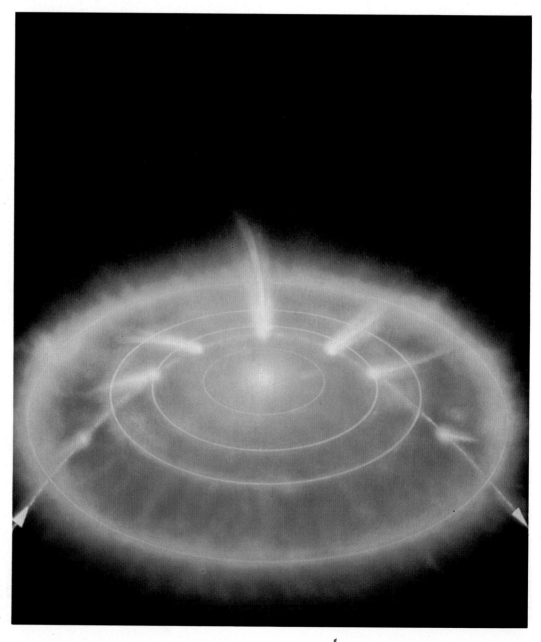

Far right: Neptune's Great Dark Spot.

APPENDIX I   SUCCESSFUL PLANETARY AND COMETARY PROBES

Planet	Probe	Date of Launch	Date of Encounter	Notes
MERCURY	Mariner 10 (USA)	3 November 1973	29 March 1974 21 September 1974 16 March 1975	First (and so far only) Mercury probe; first multiple planet swing-by
VENUS	Mariner 2 (USA)	27 August 1962	14 December 1962	First successful Venus flyby
	Venera 4 (USSR)	12 June 1967	18 October 1967	First successful descent through Venusian atmosphere; data transmitted for 94 minutes during descent
	Mariner 5 (USA)	14 June 1967	19 October 1967	Second flyby of Venus by a US probe
	Venera 5 (USSR)	5 January 1969	16 May 1969	Second successful descent through Venusian atmosphere; data transmitted for 53 minutes during descent
	Venera 6 (USSR)	10 January 1969	17 May 1969	Third successful descent through Venusian atmosphere; data transmitted for 51 minutes during descent
	Venera 7 (USSR)	17 August 1970	15 December 1970	First probe to be still transmitting after reaching Venusian surface; data transmitted for 23 minutes after landing
	Venera 8 (USSR)	27 March 1972	22 July 1972	Second probe to transmit from Venusian surface; transmitted for 50 minutes after landing
	Mariner 10 (USA)	3 November 1973	5 February 1974	Transmitted pictures of Venusian cloud tops; went on to a triple flyby of Mercury
	Venera 9 (USSR)	8 June 1975	22 October 1975	First picture from Venusian surface; transmitted for 53 minutes after landing
	Venera 10 (USSR)	14 June 1975	25 October 1975	Second picture from Venusian surface; transmitted for 65 minutes after landing
	Pioneer Venus 1 (USA)	20 May 1978	4 December 1978	First radar maps of Venusian surface
	Pioneer Venus 2 (USA)	8 August 1978	9 December 1978	Multi-probe mission; first US atmospheric probe to Venus
	Venera 11 (USSR)	9 September 1978	25 December 1978	Sixth Venus landing; transmitted for 95 minutes after landing
	Venera 12 (USSR)	14 September 1978	21 December 1978	Fifth Venus landing; transmitted for 110 minutes after landing
	Venera 13 (USSR)	30 October 1981	1 March 1982	First colour picture from Venusian surface; first soil analysis
	Venera 14 (USSR)	4 November 1981	5 March 1982	Second colour picture; second soil analysis
	Venera 15 (USSR)	2 June 1983	10 October 1983	Second Venus radar mapper (first from USSR)
	Venera 16 (USSR)	7 June 1983	14 October 1983	Third Venus radar mapper (second from USSR)
	Vega 1 (USSR)	15 December 1984	11 June 1985	First balloons dropped into Venusian atmosphere; went on to a flyby of Halley's Comet
	Vega 2 (USSR)	21 December 1984	15 June 1985	Second mission involving the deployment of balloons into Venusian atmosphere; went on to a flyby of Halley's Comet

MARS	Mariner 4 (USA)	28 November 1964	15 July 1965	First Mars flyby; first images of Mars (21 received)
	Mariner 6 (USA)	25 February 1969	31 July 1969	Second successful Mars flyby
	Mariner 7 (USA)	27 March 1969	5 August 1969	Third successful Mars flyby
	Mars 2 (USA)	19 May 1971	27 November 1971	First Mars lander (failed); data transmitted by orbiter
	Mars 3 (USA)	28 May 1971	2 December 1971	Second Mars lander (failed after 20 seconds); data transmitted by orbiter
	Mariner 9 (USA)	30 May 1971	14 November 1971	First Martian artificial satellite; transmitted 7,293 pictures including images of Phobos and Deimos
	Viking 1 (USA)	20 August 1975	19 June 1976	First successful Mars lander; first photographs from surface
	Viking 2 (USA)	9 September 1975	7 August 1976	Second successful Mars lander
JUPITER	Pioneer 10 (USA)	3 March 1972	4 December 1973	First flyby of Jupiter; first probe to eventually leave Solar System; first crossing of asteroid belt
	Pioneer 11 (USA)	6 April 1973	3 December 1974	Successful follow-up to Pioneer 10
	Voyager 1 (USA)	5 September 1977	5 March 1979	Good images of Jupiter and the Galilean satelites Io, Ganymede and Callisto; discovery of ring system and volcanoes on Io
	Voyager 2 (USA)	20 August 1977	9 July 1979	Successful follow-up to Voyager 1; high resolution images of Europa; good images of ring system and volcanoes on Io
SATURN	Pioneer 11 (USA)	6 April 1973	1 September 1979	First flyby of Saturn
	Voyager 1 (USA)	5 September 1977	12 November 1980	Good images of Titan, Rhea and Mimas
	Voyager 2 (USA)	20 August 1977	26 August 1981	Good images of Hyperion, Tethys and Enceladus
URANUS	Voyager 2 (USA)	20 August 1977	24 January 1986	First fly-by of Uranus; ten new satellites discovered; number of known rings increased to eleven
NEPTUNE	Voyager 2 (USA)	20 August 1977	25 August 1989	First flyby of Neptune; six new satellites discovered; presence of ring system confirmed
HALLEY'S COMET	Vega 1 (USSR)	15 December 1984	6 March 1986	First close flyby of Halley's Comet
	Vega 2 (USSR)	21 December 1984	9 March 1986	Second close flyby of Halley's Comet
	Sakigake (Japan)	7 January 1985	11 March 1986	First Japanese deep-space probe
	Giotto (European)	2 July 1985	14 March 1986	First close encounter with a comet; good images of nucleus obtained; first European deep-space project
	Suisei (Japan)	18 August 1985	8 March 1986	Second Japanese deep-space probe

APPENDIX 2   TABLE OF PLANETARY DATA

Planet	Equatorial Diam (km)	Distance From Sun (km)	Sidereal Period (Year)		Axial Rotation Period (Equatorial)	
Mercury	4,880	57,900,000	87.97	days	58.65	days
Venus	12,104	108,200,000	224.7	days	243.01	days
Earth	12,756	149,600,000	365.265	days	23.93	hours
Mars	6,787	227,900,000	686.98	days	24.62	hours
Jupiter	143,800	778,300,000	11.86	years	9.84	hours
Saturn	120,660	1,427,000,000	29.46	years	10.23	hours
Uranus	50,800	2,869,600,000	84.01	years	17.24	hours
Neptune	49,500	4,496,600,000	164.79	years	17.80	hours
Pluto	2,300	5,900,100,000	248	years	6.3874	days

	Axial Tilt (°)	Inclination of Orbit to Ecliptic (°)	Orbital Eccentricity	Mean Density (g/cm³)
Mercury	0.0	7.0	0.206	5.42
Venus	177.3	3.3	0.007	5.25
Earth	23.45	0.0	0.017	5.52
Mars	23.99	1.9	0.093	3.94
Jupiter	3.13	1.3	0.048	1.31
Saturn	29.0	2.5	0.056	0.69
Uranus	97.9	0.77	0.047	1.3
Neptune	28.8	1.77	0.009	1.66
Pluto	118.0	17.2	0.250	1.8

	Mass (Earth = 1)	Maximum Magnitude	Escape Velocity (km/sec)	Number of Satellites
Mercury	0.06	−1.9	4.3	0
Venus	0.86	−4.4	10.4	0
Earth	1.00	—	11.2	1
Mars	0.11	−2.8	5.0	2
Jupiter	317.89	−2.6	59.6	16
Saturn	95.15	−0.3	35.6	21
Uranus	14.54	+5.6	21.2	15
Neptune	17.23	+7.7	23.6	8
Pluto	0.002	+14	1	1

(To convert kilometres to miles, divide by 1.6)

Glossary

APHELION The point in its ORBIT around the SUN at which an object is furthest from the SUN.

APOGEE The point in its ORBIT around the Earth at which an object is furthest from the Earth.

ASHEN LIGHT A dim glow sometimes seen on the dark side of Venus when it is visible as a thin crescent. The cause is not fully understood.

ASTEROID Another name for MINOR PLANET.

AURORA Glows seen over the polar regions which occur when energized particles from the SUN react with particles in the Earth's upper atmosphere.

AUTUMNAL EQUINOX The point at which the apparent path of the SUN, moving from north to south, crosses the CELESTIAL EQUATOR.

CELESTIAL EQUATOR A projection of the Earth's equator onto the CELESTIAL SPHERE, equidistant from the CELESTIAL POLES and dividing the CELESTIAL SPHERE into two hemispheres.

CELESTIAL POLES The points on the CELESTIAL SPHERE directly above the north and south terrestrial poles around which the CELESTIAL SPHERE appears to rotate.

CELESTIAL SPHERE The imaginary sphere of STARS surrounding the Earth.

COMET An object comprised of a mixture of gas, dust and ice and which travels around the SUN in an ORBIT that is usually very eccentric.

CONJUNCTION The position at which two objects are lined up with each other (or nearly so) as seen from Earth. Superior conjunction occurs when a PLANET is at the opposite side of the SUN as seen from Earth and inferior conjunction when a PLANET lies between the SUN and EARTH.

ECLIPSE The obscuration of one celestial object by another, such as the SUN by the MOON during a solar eclipse.

ECLIPTIC The apparent path of the SUN through the sky. The ecliptic passes through a band of CONSTELLATIONS called the Zodiac.

ELLIPSE The closed, oval-shaped form obtained by cutting through a cone at an angle to the main axis of the cone. The orbits of the PLANETS around the SUN are all elliptical.

EQUINOX The AUTUMNAL and VERNAL equinoxes are the two points at which the ECLIPTIC crosses the CELESTIAL EQUATOR.

FIREBALL A very bright METEOR or SHOOTING STAR.

INFERIOR PLANET A planet that travels around the SUN inside the ORBIT of the Earth.

LIMB The edge of the visible disc of an object such as the MOON or a PLANET.

MERIDIAN An imaginary line crossing the CELESTIAL SPHERE and which passes through both CELESTIAL POLES and the ZENITH.

METEOR A streak of light in the sky seen as the result of the destruction through atmospheric friction of a METEOROID in the Earth's atmosphere.

METEORITE A METEOROID which is sufficiently large to at least partially survive the fall through Earth's atmosphere.

METEOROID A term applied to particles of interplanetary meteoritic debris.

MINOR PLANET One of a large number of small planetary bodies which orbit the SUN largely between the ORBITS of Mars and Jupiter. Also called ASTEROIDS.

MOON The Earth's only natural SATELLITE.

NADIR The point on the CELESTIAL SPHERE directly opposite the ZENITH.

OCCULTATION The temporary covering up of one celestial object, such as a STAR, by another, such as the MOON.

OPPOSITION The point in its ORBIT at which a SUPERIOR PLANET is directly opposite the SUN in the sky.

ORBIT The closed path of one object around another.

PENUMBRA The lighter part of a sunspot. Also the area of partial shadow around the main cone of shadow cast by the MOON during a solar ECLIPSE or the Earth during a lunar ECLIPSE.

PERIGEE The point in its ORBIT around the Earth at which an object is closest to the Earth.

PERIHELION The point in its ORBIT around the SUN at which an object is closest to the Sun.

PLANET One of the nine major members of the SUN's family.

PRECESSION The shift of the CELESTIAL POLE and EQUINOXES caused largely by the gravitational influence of the SUN and MOON on the Earth's equatorial bulge.

PRIME MERIDIAN The MERIDIAN that passes through the VERNAL EQUINOX.

SATELLITE A small object orbiting a larger one.

SHOOTING STAR The popular name for a METEOR.

SIDEREAL PERIOD The time taken for an object to complete one ORBIT around another.

SOLAR SYSTEM The collective description given to the system dominated by the SUN and including the PLANETS, MINOR PLANETS, COMETS, planetary SATELLITES and interplanetary debris that travel in ORBITS around the SUN.

SOLAR WIND The constant stream of energized particles emitted by the SUN.

SOLSTICE The positions in the sky at which the SUN is at its maximum angular distance (Declination) from the CELESTIAL EQUATOR.

SUN The star which dominates the SOLAR SYSTEM.

SUPERIOR PLANET A PLANET that travels around the SUN outside the ORBIT of the Earth.

SYNODIC PERIOD The interval between successive OPPOSITIONS of a PLANET or other object in the SOLAR SYSTEM.

TERMINATOR The division between the light and dark hemispheres of a PLANET or SATELLITE.

TRANSIT The passage of an object across the observer's MERIDIAN or of one object across the face of another.

UMBRA The darker part of a sunspot. Also the main cone of shadow cast by the MOON during a solar ECLIPSE or the Earth during a lunar ECLIPSE.

VERNAL EQUINOX The point at which the apparent path of the SUN, moving from south to north, crosses the CELESTIAL EQUATOR.

ZENITH The point in the sky directly above the observer.

Index

Figures in *italic* refer to illustrations in the text

Adams, John Couch, 56
Adrastea, 43
Airy, George Biddell (Astronomer Royal), 56
Amalthea, 43
Amor Asteroids, 29
Ananke Carme, 44
Antoniadi, Eugenios, 17, 18
Aphrodite Terra, Venus, 20, *21*
Apollo Asteroids, 29, 30, *30*
Arethusa (asteroid), *26*
Ariel, 53, *53*
Arizona Meteorite Crater, *31*, 79
Armstrong, Neil, 185
Asteroids, 26–31
    discovery of, 26–28
    first ten discovered, 27
    formation and distribution of, 28
    naming of, 28
    orbits of, 28
    ten largest, *26*, 27
Astraea (asteroid), 27, 28

Barnard, Edward Emerson, 23, 43
Beer, Wilhelm, 22
Belt Asteroids, 29
Beta Regio, Venus, 20, 21
Biela's Comet, 68
Biot, Jean Baptiste, 78
Bode, Johann, 26
Bode's Law, 26
Brahe, Tycho, 8, *9*
Brucia, 28

Callisto, 34, 40, *40*, 41, 42, *43*, 91
Caloris Basin, Mercury, 17, *18*
Cape Kennedy, 34
Carrington, Richard, 14
Cassini Division, 49, *49*
Cassini, Giovanni Domenico, 22, 48
Cassini space mission, 32, 47, 87, *87*
'Celestial Police', 27
Celestial Sphere, 6, *6*, 7
Ceres (asteroid), *26*, 27
Charon, 62–63, *63*
China (asteroid), 28
Christy, James W., 62
chromosphere, 14
Cleopatra Patera crater, Venus, 21
Collegio Romano Observatory, 69
Comet Grigg-Skjellerup, 73, 88

Comet Kobayashi-Berger-Milon, *68*
Comet-Nucleus-Sample-Return (CNSR) mission, 81
Comet Rendezvous and Asteroid Flyby (CRAF), *74*, 75
Comet Tempel-Tuttle, 77
Comet West, *64*
cometary orbits, *66*, 67
comets, 64–75
    structure of, *65*
constellations, 6
Copernican Theory, 7, 8
Copernicus, Nicolaus, 7
Cybele (asteroid), 27

Dark Spot 2, Neptune, 57, *58*
d'Arrest, Heinrich, 56
Davida (asteroid), 27
Deep Space Network (DSN), 35–36
Deimos, 24, *24*, 25, *25*, 91
Diana Chasma, Venus, 21
differential rotation, 14
Dione, *46*, 48
Doppler Effect, 19

ecliptic, 16
Einstein (asteroid), 28
Elara, 44
Enceladus, *45*, *46*, 47, *48*, 91
Encke Division, 49
Encke, Johann, 49, 67
Encke's Comet, 67, 77
Epsilon ring, Uranus, 55, *56*
Eros (asteroid), 28, 30, *30*, 31
Euphrosyne (asteroid), 27
Europa (asteroid), 27, 34, 40, *40*, 41, 42, 91
European Space Agency (ESA), 81, 87, *87*, 88

faculae, 14
Flora (asteroid), *26*, 27

Galileo, Galilei, 8, 40, 49, 81
Galileo probe, 10, 32, 38, 81, *83*, 85, *85*
Galileo Regio, Ganymede, 41
Galle, Johann, 56
Gamma Leonis, 78
Ganymede, 34, 40, *40*, 42, 91
Gaspra (asteroid), 81
Gauss, Carl Friedrich, 27
Giotto probe, *71*, *72*, 73, 88, 91
granulation, 14
Great Dark Spot, Neptune, 56, 57, *58*, *89*
Great Daylight Comet, 69
Great Red Spot, Jupiter, 37, 38, *38*, 39

ACKNOWLEDGMENTS

Annabelle Trodd: 40 top.
Australian Overseas Information Service, London: 78.
Brian Trodd Publishing House: 22 top.
Don Dixon: 24 top.
ESA: 16, 29, 36, 65 bottom, 67 top, 71, 72, 73, 87.
NASA: 4, 5, 10 bottom, 14, 15 top & bottom, 17, 18 top & bottom, 19 top & bottom, 21 top & bottom, 22 bottom, 23, 25, 33, 34, 35, 36, 37, 38, 39, 40, 41, 42, 43 bottom, 44, 45, 46, 47, 48, 49, 50, 51, 52, 53, 54, 55, 56, 57, 58, 59 top & bottom, 60, 61, 67 bottom, 74, 75, 80, 82, 83, 84/85, 85 bottom, 86, 87, 89.
Nicholas Booth: 24 bottom, 43 top, 70.
Novosti: 20.
Patrick Moore Collection: 8 bottom, 31, 79
Paul Doherty: 2, 3, 11, 12, 13 top & bottom, 26, 30 top & bottom, 62/63 top, 65 top, 66, 76, 77.
Science Photo Library/Dr. Jeremy Burgess: 8 top.
Science Photo Library/J.-L. Charmet: 9.
Science Photo Library/Dennis Milon/John Labord: 68.
Science Photo Library/Ronald Royer: 64, 69.
Science Photo Library/Julian Baum: 88.
Scientific American: 32.

Grigg, John, 88

Halawe (asteroid), 28
Hall, Asaph, 24, 25
Hall crater, Mars, 25
Halley, Edmund, 68
Halley's Comet, 10, 65, 67, 67, 68–75, 69, 70, 72, 73, 75, 88, 88, 90, 91
Harding, Karl, 27
Hayden Planetarium, New York, 79
Hebe (asteroid), 27, 28
heliosphere, 15, 16
Hencke, Carl, 28
Herculina (asteroid), 31
Hermes (asteroid), 30
Herschel crater, Mimas, 47, 48
Herschel, William, 22, 24, 26, 50, 52, 56
Hidalgo (asteroid), 26, 30
Himalia, 44
Hoba West Meteorite, S. Africa, 79
Huygens, Christiaan, 22, 47, 49
Hygeia (asteroid), 27
Hyperion, 46, 91

Iapetus, 46, 48
Ibsen crater, Mercury, 17
Ida (asteroid), 81
Interamnia (asteroid), 27
International Solar Polar Mission (ISPM), 16
Io, 10, 34, 40, 40, 41, 41, 42, 85, 91
Iris (asteroid), 26, 27
Ishtar Terra, Venus, 20, 21
Ithaca Chasma, Tethys, 48

Juno (asteroid), 27
Jupiter, 7, 10, 16, 26, 28, 29, 32, 34, 37, 37–44, 45, 46, 63, 81, 83, 85, 91, 92
   rings of, 43, 44, 44
   satellites of, 40–44

Kepler, Johannes, 8, 8
Kirkwood, Daniel, 29
Kirkwood Gaps, 29
Kuiper, Gerard, 47

Lac Couture crater, Quebec, 79
Lagrange points, 29
Le Verrier, Urbain J.J., 56
Leonids, 78
Lesa, 44
Lowell Observatory, Arizona, 62
Lowell, Percival, 22, 62, 63
Lysithea, 44

Magellan probe, 85, 85, 86, 86
Maja (asteroid), 87

Manson crater, Iowa, 79
Mariner probes, 8, 17, 18, 23, 24, 25, 81, 90, 91
Mars, 7, 8, 10, 20, 21, 22–25, 26, 28, 29, 81, 91, 92
Mars Observer mission, 81
Mars Rover Sample Return Mission, 81
Maxwell Montes, Venus, 21
Mercury, 6, 7, 8, 17–18, 17, 19, 20, 31, 81, 90, 92
meteors and meteorites, 77–79
Metis (asteroids), 27, 43
Milky Way, 6
Mimas, 34, 45, 46, 47, 47, 91
Miranda, 8, 52, 52

N1 satellite, 59, 60, 61
NASA, 34, 35, 75, 87, 87
Neptune, 7, 8, 10, 32, 35, 36, 56–61, 57, 59, 62, 81, 91, 92
   rings of, 60–61, 61
   satellites of, 57–61
Nereid, 59–60
NORC (asteroid), 28
Nyasa (asteroid), 26

Oberon, 52
Odysseus crater, Tethys, 48
Olbers, Heinrich, 27
Olympus Mons, Mars, 10, 23
Oort, Jan, 65
Oort's Cloud, 65, 66

Palitzsch, Johann Georg, 68
Pallas (asteroid), 27, 28
Parkes Radio Observatory, Australia, 36
Pasiphae, 44
Patientia (asteroid), 27
Payne-Gaposchkin (asteroid), 28
Petrarch crater, Mercury, 17
Phobos, 24, 25, 25, 91
Phoebe, 46, 48
photosphere, 13, 15
Piazzi, Giuseppe, 27
Pioneer probes, 32, 36, 37, 38, 39, 44, 47, 49, 81, 90, 91
planetary data, in detail, 91
planetary exploration, 80–88
Pluto, 6, 7, 10, 36, 62, 63, 65, 81, 92
protostar, 11, 12
Ptolemaic Theory, 7, 8, 8
Ptolemy, Claudius, 7, 7
Puck, 53

Rhea, 34, 46, 47, 48, 91
Rheas Mons, Venus, 21

Sakigake probe, 72, 91
Saturn, 7, 10, 32, 34, 44, 45–49, *45*, 81,
    87, 91, 92
  rings of, 49
  satellites of, *46*, 47–49, *47*
Schiaparelli, Giovanni Virginio, 17, 22
Schröter, Johann H., 19, 26
Schwabe, Heinrich, S., 13
Sinope, 44
Skjellerup, J.F., 88
Solar corona, 14, 15, *15*
Space Shuttle, 16, *83*, 85, *85*
Stickney crater, Mars, 25
Suisei probe, 72, 91
Sun, the, 13–15, 16
  flares, 14, 15, *15*
  prominences, 14, 15
  structure, *13*
  sunspots, 13, 14, *14*
Swift, Jonathan, 24, 25
Syrtis Major plateau, Mars, 22

Taurids, 77, 78
Tethys, 34, *45*, *46*, 47, *47*, 91
Tharsis region, Mars, 23
Thebe, 43
Theia Mons, Venus, 21
Titan, 10, 32, 34, 45, *46*, 47, 87, 91
Titan probe, *87*
Titania, *51*, 52
Titius, Johann, 26, 27
Tombaugh, Clyde, 7, 62
Triton, 8, 57–59, *59*, 60, *60*

Trojan Asteroids, 28, 29

Ulysses mission, 16, *16*
Uranus, 7, 26, 32, 34, 36, 44, 50–56, *50*,
    69, 81, 91, 92
  rings of, 54–55, *54*, *55*
  satellites of, 52–54, 56
US Naval Observatory, Washington
    DC, 24

Valhalla Basin, Callisto, 42
Vallis Arecabo, Mercury, 17
Vallis Marineris, Mars, 21, *23*, 24
Vega probes, 70, 73, 90, 91
Venera probes, 8, 20, *20*, 90
Venus, 7, 8, 19–21, *19*, 70, 81, 85, 86,
    90, 92
Very Large Array (VLA), 36
Vesta (asteroid), *26*, 27, 28
Viking probes, *22*, 23, 24, 81, 91
Voltaire, Francois-Marie, 24, 25
von Madler, Johann, 22
von Zach, Baron Franz, 27
Voyager probes, 8, 10, *10*, 32–63, *32*, *33*,
    34, 81, 91

Wolf, J. Rudolf, 13
Wolf, Max, 28, 69
Wolfe Meteorite Crater, W. Australia,
    78

Zodiac, the, 6